MATERIALS SCIENCE AND TECHNOLOGIES

PLASTICITY OF CRYSTALS IN A MICROSCOPIC VIEWPOINT ON THE BASIS OF DISLOCATION MOTION

MATERIALS SCIENCE AND TECHNOLOGIES

Additional books in this series can be found on Nova's website under the Series tab.

Additional E-books in this series can be found on Nova's website under the E-book tab.

MATERIALS SCIENCE AND TECHNOLOGIES

PLASTICITY OF CRYSTALS IN A MICROSCOPIC VIEWPOINT ON THE BASIS OF DISLOCATION MOTION

YOHICHI KOIIZUKI

Nova Science Publishers, Inc.
New York

For permission to use material from this book please contact us:
Telephone 631-231-7269; Fax 631-231-8175
Web Site: http://www.novapublishers.com

Library of Congress Cataloging-in-Publication Data

Plasticity of crystals in a microscopic viewpoint on the basis of dislocation motion / [edited by] Yohichi Kohzuki.
p. cm.
Includes bibliographical references and index.
ISBN 978-1-62081-419-2 (hardcover)
1. Dislocations in crystals. 2. Crystals--Defects. I. Kohzuki, Yohichi.
QD945.P54 2011
548'.842--dc23
2012011773

Published by Nova Science Publishers, Inc. † New York

CONTENTS

Preface **xi**

Chapter 1 Process for Thermal Activated Overcoming of the Short-Range Obstacle by a Dislocation **1**

1.1. Hardening of Alkali Halide Crystals Due to the Different Types of Atomic Defects **1**

1.2. Common Type of Thermal Activation Barrier for a Dislocation Motion **2**

References **6**

Chapter 2 Strain-Rate Cycling Tests Associated with Ultrasonic Oscillation **9**

2.1. Introduction **9**

2.2. Experimental Procedure **13**

2.3. Preparation of Specimens **15**

2.4. $\Delta\tau'/\Delta Ln\,\dot{\varepsilon}$ Versus $\Delta\tau$ for Various Specimens **17**

2.5. τ_{P1} And τ_{P2} Versus Temperature for $Kcl:Sr^{2+}$ **21**

Single Crystals **21**

2.6. τ_{P1} And τ_{P2} Versus Yield Stress for $Kcl:Sr^{2+}$ **22**

Single Crystals **22**

2.7. Discussion for the Relation between the Strain-Rate Sensitivity and the Stress Decrement for $Kcl:Sr^{2+}$ **23**

Single Crystals **23**

Conclusion **25**

References **25**

Chapter 3 Influence of State of Impurities on the Effective Stress Due to the Impurities **29**

3.1. Introduction **29**

3.2. Relation between the Strain-Rate Sensitivity and the Stress Decrement for the Two Kinds of Specimens **30**

3.3 Dependence of the State of Impurities for $Kcl:Sr^{2+}$ on Heat Treatment **31**

3.4. Effective Stress and Critical Temperature **33**

Conclusion **36**

References **36**

Chapter 4	Effective Stress and Critical Temperature of KCl Doped with Various Divalent Cations	**37**
	4.1. Introduction	**37**
	4.2. Relation between the Temperature and τ_{P1}, τ_{P2}, and Yield Stress τ_Y	**37**
	4.3. Critical Temperature	**42**
	4.4. Relation between the Strain-Rate Sensitivity and the Stress Decrement	**47**
	4.5. Activation Energy for the Interaction between a Dislocation and the Divalent Impurity	**50**
	Conclusion	**53**
	References	**54**
Chapter 5	The Friedel Relation and Thermally Activated Flow Parameters of KCl Doped with Various Divalent Cations	**55**
	5.1. Introduction	**55**
	5.2. Relation between $(L_0/L)^2$ and Φ_0	**55**
	5.3. Relation between Effective Shear	**59**
	Stress and Temperature	**59**
	5.4. Bending Angle of a Dislocation	**63**
	Pinned by the Impurity	**63**
	5.5. Dependence of the Strain-Rate Sensitivity Due to the Impurities on the Temperature	**64**
	5.6. The Enthalpy and the Gibbs Free Energy of Activation for the Break-Away of a Dislocation from the Various Divalent Impurities	**66**
	Conclusion	**71**
	References	**72**
Chapter 6	Applicability of the Friedel Relation to the Interaction between a Dislocation and the Impurity in KCl: Mg^{2+} Single Crystals	**73**
	6.1. Introduction	**73**
	6.2. Relation between Effective Shear	**73**
	Stress and Temperature	**73**
	6.3. Dependence of the Strain-Rate Sensitivity Due to the Impurities on the Temperature	**75**
	6.4. Relation between the Effective Shear Stress and the Activation Volume for the Overcoming of the	**78**
	Impurities by a Dislocation	**78**
	6.5. Consideration with $m_{\varepsilon=0}$	**80**
	6.6. Relation between Temperature and the Activation Enthalpy for the Overcoming of the	**85**
	Impurities by a Dislocation	**85**
	6.7. Consideration with $\tau_{P1}(\Delta Ln\,\dot{\varepsilon}/\Delta\tau')_P$	**87**

6.8. Gibbs Free Energy for Overcoming of Tetragonal Lattice Distortions by a Dislocation 90
Conclusion 92
References 93

Chapter 7 Influence of State of Impurities on the Deformation Characteristics 95
7.1. Introduction 95
7.2. Discussion for the Force-Distance Relation between a Dislocation and the Aggregate 97
7.2.1. Relationship of τ_{P1} and Temperature 97
7.2.2. Relative Curve of Temperature and the Strain-Rate Sensitivity due to Aggregates 99
7.2.3. Dislocation Velocity-Effective Stress Exponent 101
7.2.3.1. Quenched Specimen 101
7.2.3.2. Annealed Specimen 104
7.2.4. F_0, L/L_0, and Φ_0 108
*7.2.5. Influence of Heat Treatment on m** 110
7.2.5.1. m Obtained on the Basis of $m_{\varepsilon=0}$* 110
7.2.5.2. m Obtained on the Basis of the Relation between Strain-Rate Sensitivity and $\Delta\tau$* 111
7.2.6. Proportionality of $\Delta H(T)$ 112
7.3. The Gibbs Free Energy for Overcoming of the Aggregate by a Dislocation 115
7.4. Influence of the State of Impurities Due to the 119
Heat Treatment on the Activation Energy for 119
Breakaway from the Impurity by a Dislocation 119
7.4.1. Activation Enthalpy for the Dislocation Motion 119
7.4.2. Gibbs Free Energy for the Dislocation Motion 121
Conclusion 122
References 123

Chapter 8 Effective Stress and Critical Temperature of KCl Doped with Monovalent Cations 125
8.1. Introduction 125
8.2. Relation between the Strain-Rate Sensitivity and the Stress Decrement 125
8.3 Influence of Temperature on the Strain-Rate Sensitivity Versus Stress Decrement Curve 128
8.4. Dependence of τ_{P1}, τ_{P2}, and τ_Y on the Temperature 131
8.5. Interpretation of Strain-Rate Sensitivity Versus Stress Decrement Curve 131
8.6. Relation between the Effective Stress Due to Impurities and Temperature 133
8.7. Dependence of the Strain-Rate Sensitivity Due to the Impurities on the Temperature 139

8.8. Relation between Temperature and the Activation Enthalpy for the Interaction between a **142**
Dislocation and the Impurity *142*
8.9. Gibbs Free Energy for the Breakaway of a **143**
Dislocation from the Impurity **143**
Conclusion **145**
References **145**

Chapter 9 Bending Angle at Which the Dislocation Breaks Away from the Obstacle in KCl Doped with Various Monovalent Impurities **147**
9.1. Introduction **147**
9.2. Theoretical Concept **148**
9.3. Bending Angle **149**
Conclusion **152**
References **152**

Chapter 10 Influence of a Strain on the Interaction between a Dislocation and the Impurities **155**
10.1. Introduction **155**
10.2. Influence of Strain on the Strain-Rate Sensitivity Versus Stress Decrement Curve **155**
10.2.1. Relation between the Strain-Rate Sensitivity and the Stress Decrement **155**
10.2.2. Dependence of τ_{P1} and τ_{P2} on the Shear Strain **158**
10.3. Critical Temperature and the Effective Stress **159**
10.4. Strain-Rate Sensitivity at the Second Plateau **159**
Place on the Strain-Rate Sensitivity Versus Stress Decrement Curve **159**
Conclusion **163**
References **164**

Chapter 11 Influence of the Divalent Impurities on Dislocation Density of KCl Doped with Various Divalent Impurities **165**
11.1. Introduction **165**
11.2. Influence of the Divalent Impurities on the Forest Dislocation Density **166**
11.2.1. Influence of the Divalent Impurities on the Forest Dislocation Density in Different Plastic Deformation Regions of Stress versus Strain Curve **166**
11.2.2. Influence of the Concentration of Divalent Impurities on the Forest Dislocation Density **170**
11.2.3. Influence of the Size of Divalent Impurity on the Forest Dislocation Density **171**
11.3. Influence of the Divalent Impurities on Mobile Dislocation Density **172**
Conclusion **175**

References **176**

Chapter 12 Influence of the Heat Treatment on Deformation Characteristics **177**

12.1. Introduction **177**

12.2. Influence of the Heat Treatment on Moving Dislocation Density **177**

12.3. Strain-Rate Sensitivity at Second Plateau Place **182**

on the Relative Curve of Strain-Rate Sensitivity and **182**

Stress Decrement **182**

Conclusion **185**

References **186**

Index **187**

PREFACE

Plasticity of crystals in a microscopic viewpoint, especially on the basis of dislocation (linear defects in crystal) motion, is detailed from the data obtained mainly by the strain-rate cycling tests associated with ultrasonic oscillation (see sec. 2.2 of chap. 2 in this book) for several kinds of alkali-halide single crystals at low temperatures. The contents of this book consist of twelve chapters as below.

Chapter 1 - Process for thermal activated overcoming of the short-range obstacle (thermal obstacle) by a dislocation is presented in this chapter. Various types of barriers for a dislocation motion are introduced here. And further, a brief description of the separation of various models between a dislocation and an atomic defect is given in terms of the impurity concentration versus thermal component of flow stress (τ^*) at the temperature of 0 K and τ^* versus temperature.

Chapter 2 - The information on the interaction between a dislocation and obstacles has been provided by various methods so far. In especially, the strain-rate cycling tests associated with the ultrasonic oscillation can separate the contributions arising from the interaction between a dislocation and impurities and from the dislocations themselves during plastic deformation. This chapter presents the process of the measurement of strain-rate sensitivity under the application of ultrasonic oscillatory stress in detail.

The plots of the strain-rate sensitivity and the stress decrement due to oscillation has two bending points and two plateau regions for $KCl:Sr^{2+}$ and seems to reflect the influence of ultrasonic oscillation on the dislocation motion on the slip plane. The first bending point, τ_{p1}, at low stress decrement corresponds to the effective stress due to weak obstacles on the mobile dislocation.

Chapter 3 - If the I-V dipoles in $KCl:Sr^{2+}$ single crystals diffuse into the matrix, or they aggregate by heat treatment, various deformation characteristics will be changed. This is because the state of a small amount of impurities (I-V dipoles) in the crystals strongly influences the resistance to movement of a dislocation. In this chapter, influence of a state of impurities on the effective stress due to the impurities is described for $KCl:Sr^{2+}$ single crystals.

Chapter 4 - It is presented whether the size of divalent cations and its state are important factors for the solution hardening of KCl: Mg^{2+}, Ca^{2+}, Sr^{2+} or Ba^{2+} single crystals. The main aim of chapter 4 is to estimate the effect of various divalent cations on the critical temperature (T_c), at which the effective stress due to a weak obstacle lying on the dislocation is zero, and on the activation enthalpy $\Delta H(T_c)$ for the break-away of a dislocation from the dopant cation.

Chapter 5 - It is well known that the Friedel relation between the effective stress and the average length of dislocation segments can be applied to most weak obstacles to dislocation motion at low solute concentration. This chapter presents the applicable range of the Friedel relation on the basis of versus the bending angle of dislocation by the weak obstacle. L_0 is the average spacing of obstacles distributed randomly on the slip plane and L is the average spacing of obstacles along the dislocation. Whether the Friedel relation can be taken into the Fleischer's model or not is examined for the interaction between a dislocation and the divelent impurity in KCl: Mg^{2+}, Ca^{2+}, Sr^{2+} or Ba^{2+} single crystals. Furthermore, a difference in divalent cationic size has an influential factor of the thermally activated flow parameters: T_c, the Gibbs free energy (ΔG_0) for overcoming the impurity by a dislocation, and the tetragonality ($\Delta\varepsilon$) produced around the I-V dipoles.

Chapter 6 - It is investigated in this chapter 6 whether the Friedel relation is valid for the interaction between a dislocation and an impurity in KCl:Mg^{2+} single crystal, which is expected to cause the "rapid hardening" described in chapter 1. This is carried out from examining various relationships (e.g., effective stress versus temperature, strain-rate sensitivity due to the impurities versus temperature, effective stress versus activation volume, dislocation velocity-effective stress exponent versus temperature, activation enthalpy versus temperature, etc.). In addition, the activation enthalpy and the Gibbs free energy for overcoming the impurity by a dislocation are obtained here.

Chapter 7 - After the heat treatment, the change in deformation characteristics (e.g., T_c, the force acted on the dislocation, bending angle of dislocation by the impurity, the temperature dependence of dislocation velocity-effective stress exponent (m^*), the force-distance profile between a dislocation and the impurity, etc.) due to the agglomeration of impurities is presented in chapter 7 on the various models (a square, a parabolic, and a triangular force-distance relations between a dislocation and an aggregate and the Fleischer's model taking account of the Friedel relation) in more detail. The specimens used here are two kinds of heat-treated ones: the quenched and the annealed KCl:Sr^{2+} (0.050 mol.% in the melt) single crystals.

Chapter 8 - With respect to KCl doped with monovalent cationic impurities (Li^+ (0.50 mol.% in the melt) or Na^+ (0.50 mol.% in the melt)), various force-distance relations which represent the interaction between a dislocation and an impurity are described with a view to selecting the most suitable one from the information given by the above-mentioned method: the strain-rate cycling tests associated with the ultrasonic oscillation intermittently. The critical temperature and the activation energy for the breakaway of a dislocation from the weak obstacle such as impurity are obtained for various force-distance relations.

Chapter 9 - This chapter presents an outline of the applicable range of the Friedel relation between the effective stress and the average length of dislocation segments and refers to the applicability of the Friedel relation to the interaction between a dislocation and the monovalent impurities in various alkali-halide single crystals, which is expected to cause the gradual hardening due to the cubic symmetry defects around them.

Chapter 10 - On the basis of the variation of the strain-rate sensitivity versus stress decrement curve with shear strain, it is mainly discussed whether dislocation multiplication is influenced by Li^+ and Na^+ ions in KCl: Li^+ (0.50 mol.% in the melt) or Na^+ (0.50 mol.% in the melt) single crystals in the different plastic deformation regions (e.g., stage I, stage II, and stage III of stress-strain curve).

Chapter 11 - On the basis of the data analyzed from the variation of strain-rate sensitivity at the second plateau place on the strain-rate sensitivity versus the stress decrement curve with shear strain ($\Delta(\Delta\tau'/\Delta\ln\dot{\varepsilon})/\Delta\varepsilon$), this chapter presents whether the forest dislocation density is influenced by the concentration and the size of divalent impurities in KCl during plastic deformation. $\Delta(\Delta\tau'/\Delta\ln\dot{\varepsilon})/\Delta\varepsilon$ is considered to indicate a change in forest dislocation density with shear strain (see sec. 10.4 of chap. 10). It is also considered in more detail whether the mobile dislocation density is influenced by divalent impurities in KCl during plastic deformation, which was analyzed in terms of the relative curve of strain-rate sensitivity and stress decrement due to oscillation.

Chapter 12 - In this chapter, it is mainly investigated whether the density of the forest dislocation is changed by the heat treatment under the compression test. With respect to the two kinds of $KCl:Sr^{2+}$ (0.05 mol.% in the melt) single crystals (i.e., the quenched specimen and the annealed one), this is examined from the variation of strain-rate sensitivity at the second plateau place on the curve with shear strain ($\Delta(\Delta\tau'/\Delta\ln\dot{\varepsilon})/\Delta\varepsilon$) carried out in the preceding chapter 11. The findings will be proved by two experimental results on the strain-hardening rate and the extent of plastic deformation region of the specimens.

The preparation of specimens used here is described in the section 2.3 of chapter 2 and section 3.3 of chapter 3.

Chapter 1

PROCESS FOR THERMAL ACTIVATED OVERCOMING OF THE SHORT-RANGE OBSTACLE BY A DISLOCATION

1.1. HARDENING OF ALKALI HALIDE CRYSTALS DUE TO THE DIFFERENT TYPES OF ATOMIC DEFECTS

A large number of investigations on strength of materials have been made with alkali halide crystals so far [1-4]. This will be because the crystals have some advantages [5]: (1) simple crystal structure (i.e. rock salt structure), (2) low dislocation density in grown crystal (e.g., 10^4 cm^{-2} order for NaCl [6] and KCl [7] single crystals) as against that of annealed metals (around 10^7 cm^{-2} [8], e.g., 1 to 5×10^6 cm^{-2} in pure Cu single crystals [9], 1 to 5×10^6 cm^{-2} in α-brass single crystals [10], and 10^7 cm^{-2} order for pure Fe single crystals [11]), (3) transparency, and further (4) readily availability for pure single crystal, etc. Pure or impurity-doped alkali halide crystals, therefore, are excellent materials for an investigation on mechanical properties of crystals.

When alkali halide crystals are doped with divalent impurities (divalent cations), the impurities induce positive ion vacancies in order to conserve the electrical neutrality and are expected to be paired with the vacancies. They are often at the nearest neighbor sites forming a divalent impurity-vacancy (I-V) dipole, which attract them strongly [12], for crystals quenched from a high temperature. Then asymmetrical distortions are produced around the I-V dipoles. Mobile dislocations on a slip plane interact strongly only with these defects lying within one atom spacing of the glide plane [13]. The solution hardening is named "rapid hardening", which can be distinguished from "gradual hardening" due to the defects of cubic symmetry [14-16]. Effects of different defects on hardness of some alkali halide crystals are listed in table 1-1. The effects are expressed as increase in flow stress per the square root of concentration of point defects (i.e. $\Delta\tau/\Delta c^{1/2}$) in terms of the shear modulus, μ. Despite the same matrix (see NaCl in table 1-1), the hardening due to substitutional divalent additions is much larger than the case of monovalent ones. It has been well known for many years that aliovalent impurities (aliovalent cations) are a much more potent source of solution strengthening in ionic crystals than isovalent cations [17-19]. In view of the different types of atomic defects, solution hardening may be divided into the two classes: rapid hardening and gradual hardening by Fleischer and Hibbard [15] and Johnston et al. [16]. Roughly speaking, the value of $\Delta\tau/\Delta c^{1/2}$ for the rapid hardening is over several tens times as large as that for the gradual hardening as shown in table 1-1.

Table 1-1. Hardening of alkali halide crystals due to various defects at low concentrations below 10^{-4} [14-16]

Crystal	Different types of point defects	$\Delta\tau/\Delta c^{1/2}$ $(\times 2\sqrt{c})$ [a]
NaCl	Monovalent substitutional impurities	$< \mu/30$
KCl	F-centers, additive coloring	$\mu/2.5$
NaCl	Divalent substitutional impurities	2μ
KCl (irradiated)	Interstitial chlorine	18μ
LiF	Divalent impurity clusters	6μ
LiF	I-V dipoles (at 77 K)	10μ
LiF (irradiated)	V_K-centers (at 77 K)	25μ
LiF (irradiated)	Interstitial fluorine	5μ

[a] Δc represents the increment in the concentration of point defects and μ is the shear modulus. The measurements were made at room temperature unless otherwise noted. $\Delta\tau/\Delta c$ in references [14-16] is replaced by $\Delta\tau/\Delta c^{1/2}$.

1.2. Common Type of Thermal Activation Barrier for a Dislocation Motion

A dislocation will encounter a stress field illustrated schematically in figure 1-1 as it moves through on the slip plane containing many weak obstacles and a few strong ones. In the figure, the positive stress concerning axis of the ordinate opposes the flow stress (applied stress), τ, and the negative stress assists it. Extrinsic resistance to the dislocation motion has two types: one is long-range obstacle (the order of 10 atomic diameters or greater) and the other short-range obstacle (less than about 10 atomic diameters). The former is considered to be forest dislocations, large precipitates or second-phase particles, and grain boundary, for instance, and the latter impurity atoms, isolated and clustered point defects, small precipitates, intersecting dislocations, etc. Overcoming the latter type of obstacles (byname, thermal obstacles) by a dislocation, thermal fluctuations play an important role in aid of the flow stress above the temperature of 0 K. Then the aid energy, ΔG, supplied by the thermal fluctuations is given by the shaded part in figure 1-1. Thus the dislocation can move through below τ_0. τ_0 is the value of τ at 0 K. As for the long-range obstacles (byname, athermal obstacles), the energy barrier is so large that the thermal fluctuations play no role in overcoming them within the temperature range. If both the resistance simultaneously acts on the dislocation, the flow stress consists of two components: an athermal component (internal stress), τ_i, and a thermal component (effective stress), τ^*, which depends on temperature (T) and strain rate ($\dot{\varepsilon}$). That is,

$$\tau = \tau^*(T, \dot{\varepsilon}) + \tau_i \, . \qquad (1\text{-}1)$$

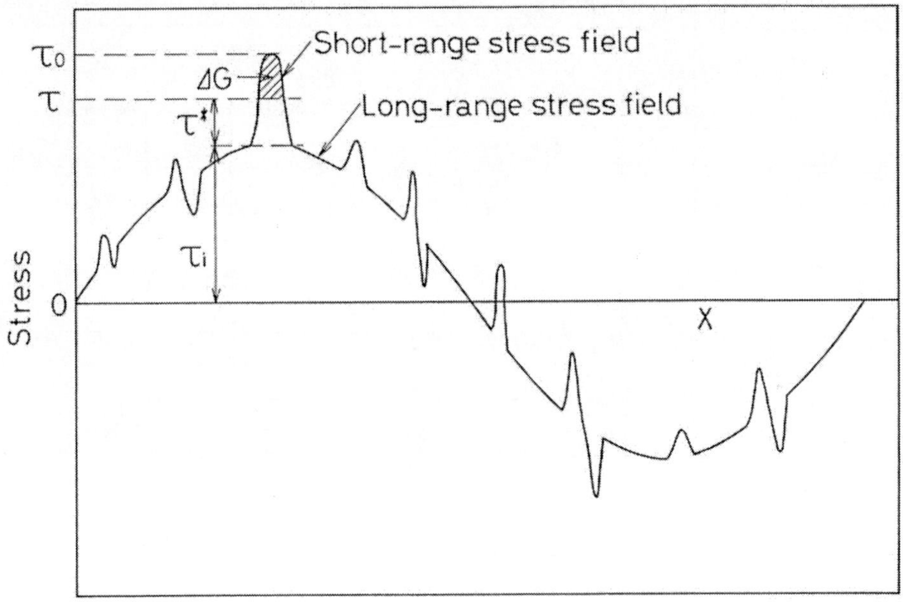

Figure 1-1. Stress fields encountered by a dislocation moving through the crystal lattice.

Relationship of the flow stress and temperature is schematically shown in figure 1-2. τ^* decreases with the temperature and becomes zero above T_c. As strain rate is increased, τ^* grows and determination of T_c is influenced by it. τ_i slightly depends on strain rate and temperature. As the magnitude of internal stress is comparable with the flow stress, the flow stress dependence of the average dislocation velocity becomes considerably high [20].

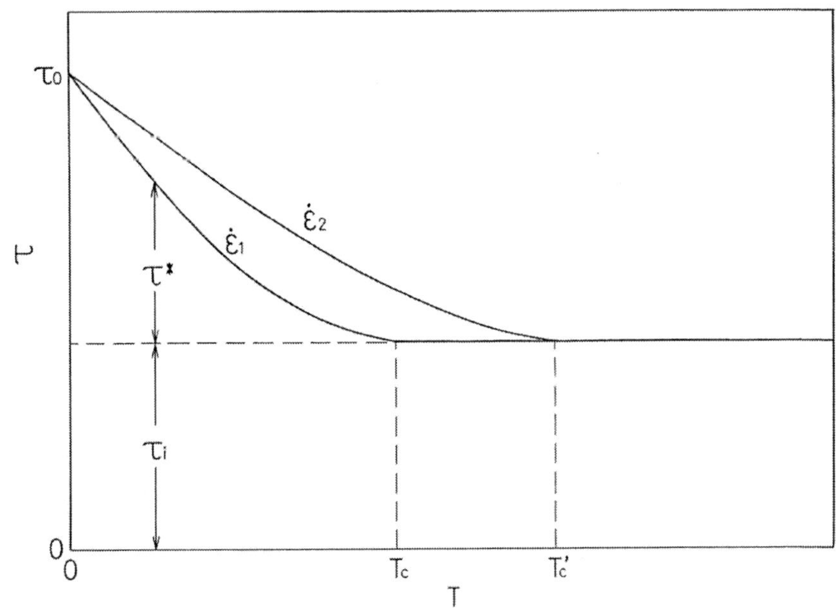

Figure 1-2. Variation of flow stress with temperature at two strain rates ($\dot{\varepsilon}_1 < \dot{\varepsilon}_2$).

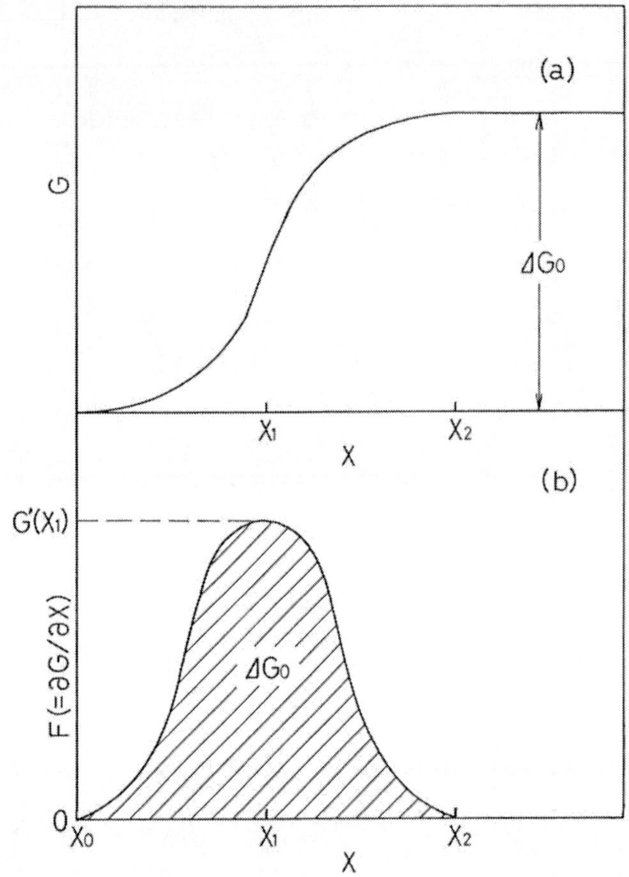

Figure 1-3. The process for thermal activated overcoming of the short-range obstacle by a dislocation. Variation in (a) the Gibbs free energy of activation and (b) the force acted on the dislocation with the distance for a dislocation motion.

The representation of figure 1-3 is concerned with a common type of thermal activation barrier. The free energy (G) varies with the distance (x) between a dislocation and the obstacle as given in figure 1-3 (a). When a dislocation overcomes the short-range obstacles, the free energy becomes high on account of the work (ΔW) done by the applied stress. Then the resistance (F), where it can be defined by the differentiation of free energy with respect to x (i.e. $\partial G/\partial x$), to the dislocation motion is revealed as figure 1-3 (b) in accord with the abscissa of figure 1-3 (a). Figure 1-3 (b) corresponds to typical force-distance curve for short-range obstacle among those in figure 1-1. Shape of this curve represented by $F(x)$ means the model overcoming the obstacle by a dislocation. ΔG_0, which is taken as the shaded area under $F(x)$ in figure 1-3 (b), is the Gibbs free energy of activation for the breakaway of the dislocation from the obstacle in the absence of an applied stress (in this case it is equivalent to the Helmholtz free energy for the dislocation motion). ΔG in figure 1-1 corresponds to the difference between ΔG_0 and ΔW, namely,

$$\Delta G = 2\int_{x_1}^{x_2}\left\{F(x) - \tau^* bL\right\}\mathrm{d}x, \quad (F(x_1) = F_0\,,\ x_2 > x_1) \tag{1-2}$$

where b is the magnitude of the Burgers vector, L is the average length of dislocation segments, x_2 is the value of x at which F equals the defect force, and F_0 is the maximum value of F (i.e. F at the temperature of 0 K).

Determining the microhardness of doped crystal, the distribution [21] and the state [22-24] of additions play an important role. Quenching a crystal from elevated temperatures, the effect of impurity concentration (c) on the flow stress is different by the state of impurities in the crystal. The variation of the thermal component of flow stress with temperature is also influenced by it. For example, it has been demonstrated that the flow stress overcoming of divalent impurities increase with $c^{1/2}$ for various alkali halide crystals (e.g., $LiF:Mg^{2+}$ at 0 K [13], $NaCl:Ca^{2+}$ at 78 K [25], $KCl:Ba^{2+}$ at room temperature [26], and single crystals of NaCl, NaBr, KCl, and KBr containing divalent additions of Ca^{2+}, Sr^{2+}, and Ba^{2+} at room temperature [22]). Then, the effective stress varies with temperature as given by

$$\left(\frac{\tau^*}{\tau_0^*}\right)^{1/2} = 1 - \left(\frac{T}{T_c}\right)^{1/2}, \quad (1\text{-}3)$$

It is investigated that the interaction between a dislocation and divalent impurities can be approximated to the Fleischer's model [13]. The Fleischer's model has been used widely with much success [13]. Many investigators concluded from the linear plots of the relation between stress to the one-half power and temperature to the one-half power that the Fleischer's model is valid, and that the obstacles have tetragonal distortion. Another several models for the force-distance profile and relation between temperature and effective stress due to the obstacle have been proposed [27, 28]. For example, the relation between the temperature and the stress is as follows

$$\left(\frac{\tau^*}{\tau_0^*}\right)^{2/3} = 1 - \left(\frac{T}{T_c}\right)^{p/(p+1)}, \quad (1\text{-}4)$$

where τ_0^* is the stress at 0 K, T_c is the temperature at which the effective stress becomes zero and a dislocation breaks away from the impurities only with the help of thermal activation, and p is a parameter. Force-distance curve is triangular when p is one [29], parabolic when p is two, and square when p is infinite [28].

As the temperature becomes higher than room temperature, reorientation of the I-V dipoles occurs due to the strain fields of moving dislocations. In this case, the flow stress increases in proportion to c but becomes independent of temperature [30]. Within the temperature range at which dislocation moves with dragging the impurities in a crystal, inverse dependence of flow stress on temperature is observed and serrated flow, which is called Portevin-LeChatelier effect [31], appears on the stress-strain curve [32, 33]. This effect is discerned for single crystals of NaCl doped with small amount of $CdCl_2$ (50 ppm) above 393 K [34]. When the temperature rises still higher, the I-V dipoles resolve in the crystal. This causes a great decrease in the flow stress.

The separation of various models between a dislocation and an impurity for a particular case has been conducted from above-mentioned two methods (i.e., c versus τ^* and T versus

τ^*) [35-38]. The models of Mott-Nabarro, Seeger, and Friedel lead to $\tau_0^* \propto c$ and the Fleischer's model $\tau_0^* \propto c^{1/2}$, provided $c > 10^{-4}$. τ_0^* is the thermal component of flow stress at 0 K. With respect to the effective stress-temperature relationship, the Mott-Nabarro model yields a straight line in the τ^*–$T^{2/3}$ plot, the Seeger's model in the $\tau^{*2/3}$–$T^{2/3}$ plot, the Friedel's model in the τ^*–T plot, the Fleischer's model in the $\tau^{*1/2}$–$T^{1/2}$ plot, the modified Fleischer's model in the $\tau^{*1/3}$–$T^{1/2}$ plot, the square potential in the $\tau^{*2/3}$–T plot, the parabolic potential in the $\tau^{*2/3}$–$T^{2/3}$ plot, and the triangular potential in the $\tau^{*2/3}$–$T^{1/2}$ plot.

REFERENCES

[1] Urusovskaya, A. A.; Darinskaya, E. V.; Voszka, R.; Jansky, J. *Cryst. Res. Technol.* 1981, *16*, 597.

[2] Suszyńska, M.; Nowak-Woźny, D. *Cryst. Res. Technol.* 1990, *25*, 855.

[3] Boyarskaya, Yu. S.; Zhitaru, R. P.; Palistrant, N. A. *Cryst. Res. Technol.* 1990, *25*, 1469.

[4] Sprackling, M. T. *The Plastic Deformation of Simple Ionic Crystals*; (edited by Alper, A. M.; Margrave, J. L.; Nowick, A. S.) Academic Press; London, 1976.

[5] Kataoka, T. *Doctoral Thesis at Osaka Univ.;* Osaka, 1975; p 2 (in Japanese).

[6] Argon, A. S.; Nigam, A. K.; Padawer, G. E. *Phil. Mag.* 1972, *25*, 1095.

[7] Kohzuki, Y. *Doctoral Thesis at Kanazawa Univ.;* Kanazawa, 1994; pp 12-13.

[8] Suzuki, H. *Introduction to Theory of Dislocations;* AGNE; Tokyo, 1989; p 70 (in Japanese).

[9] Young, Jr., F. W. *J. Appl. Phys.* 1961, *32*, 192.

[10] Meakin, J. D.; Wilsdorf, H. G. F. *Trans. Met. Soc. AIME* 1960, *218*, 737.

[11] Takeuchi, S. *J. Phys. Soc. Jap.* 1969, *27*, 929.

[12] Pick, H.; Weber, H. *Z. Phys.* 1950, *128*, 409.

[13] Fleischer, R. L. *J. Appl. Phys.* 1962, *33,* 3504.

[14] Fleischer, R. L. *Acta Metall.* 1962, *10*, 835.

[15] Fleischer, R. L.; Hibbard, W. R. *The Relation between Structure and Mechanical Properties of Metal*; Her Majesty's Stationary Office; London, 1963, p 261.

[16] Johnston, W. G.; Nadeau, J. S.; Fleischer, R. L. *J. Phys. Soc. Jap.* 1963, *Suppl.* I *18*, 7.

[17] Edner, A. *Z. Phys.* 1932, *73*, 623.

[18] Shoenfeld, H. *Z. Phys.* 1932, *75,* 442.

[19] Metag, W. *Z. Phys.* 1932, *78,* 363.

[20] Nix, W. D.; Coghlan, W.A.; Barrett,; C. R. *Mater. Sci. Eng.* 1969, *4,* 98.

[21] Lakshmipathi Rao, M.; Hari Babu, V. *Indian J. Pure Appl. Phys.* 1978, *16*, 821.

[22] Chin, G. Y.; Van Uitert, L. G.; Green, M. L.; Zydzik G. J.; Kometani, T. Y. *J. Amer. Ceram. Soc.* 1973, *56,* 369.

[23] Andreev, G. A.; Klimov, V. A. *Sov. Phys. Solid State* 1980, *22,* 2042.

[24] Narasimha Reddy, K.; Subba Rao, U. V. *Cryst. Res. Technol.* 1984, *19,* K73.

[25] Harrison, R. P.; Pratt, P. L.; Newey, C. W. A. *Proc. Brit. Ceram. Soc.* 1964, *1,* 197.

[26] Lubenets, S. V.; Startssev, V. I. *Sov. Phys. Solid State* 1968, *10*, 15.

[27] Cottrell, A. H.; Bilby, B. A. *Proc. Phys. Soc.* 1949, *A 62*, 49.

[28] Seeger, A. *Phil. Mag.* 1955, *46*, 1194.

[29] Foreman, A. J. E.; Makin, M. J. *Phil. Mag.* 1966, *14*, 911.
[30] Pratt, P. L.; Chang, R.; Newey, C. W. A. *Appl. Phys. Lett.* 1963, *3*, 83.
[31] Portevin, A.; Lechatelier, F. *Compt. Rend. Acad. Sc.* 1923, *176,* 507.
[32] Yoshinaga, H. *J. Japan Soc. for Tech. of Plasticity* 1971, *12*, 322 (in Japanese).
[33] Rodriguez, A. D.; Castaing, J. *Scripta Metall.* 1975, *9,* 551.
[34] Brown, L. M.; Pratt, P. L. *Phil. Mag.* 1963, *8*, 717.
[35] Conrad, H. *Acta Metall.* 1966, *14*, 1631.
[36] Conrad, H. *Can. J. Phys.* 1967, *45*, 581.
[37] Sprackling, M. T. *The Plastic Deformation of Simple Ionic Crystals*; (edited by Alper, A. M.; Margrave, J. L.; Nowick, A. S.) Academic Press; London, 1976; p 188.
[38] Koppenaal, T. J.; Fine, M. E. *Trans. Metall. Soc. AIME* 1962, *224*, 347.

Chapter 2

STRAIN-RATE CYCLING TESTS ASSOCIATED WITH ULTRASONIC OSCILLATION

2.1. INTRODUCTION

The interaction between a dislocation and obstacles has been investigated by measurements of yield stress [1-7] and proof stress [8, 9], microhardness test [10-14], direct observation of the dislocation [15-21], internal friction measurement [22-27], or stress relaxation test [28, 29] so far. A large number of investigations have been carried out by separation of the flow stress into effective and internal stresses on the basis of the temperature dependence of yield stress, the strain rate dependence of flow stress, and the stress relaxation. Nevertheless, it is difficult to investigate the interaction between a dislocation and obstacles during plastic deformation by the mentioned methods. This is because yield stress depends on dislocation velocity, dislocation density, and multiplication of dislocations [30] and proof stress is used as a means of assessing the yield stress. On the other hand, the effect of heat treatment on the microhardeness is almost insensitive to the change of atomic order of impurity in a specimen. As for direct observation, electron microscopy provides the information on the interaction between a dislocation and obstacles for a thin specimen but not for bulk, and also light scattering method is useful only for a transparent specimen. X-ray topography is the lack of resolution in the photograph, so that the specimen is limited to the low dislocation density below 10^4 cm^{-2}. Internal friction measurement cannot provide the information on the motion of the dislocation which moves by overcoming the forest dislocations and the weak obstacles such as impurities during plastic deformation, because the measurement concerns the motion of the dislocation which breaks away from the weak obstacles between two forest dislocations by vibration [31]. Stress relaxation test is generally assumed that internal structure of crystals does not change, i.e. dislocation density and internal stress are constant. In this book, the study on interaction between a dislocation and impurities is made by the strain-rate cycling tests during the Blaha effect measurement (see the subsequent sec. 2.2), the plasticity of crystals in a microscopic viewpoint on the basis of dislocation (linear defects in crystal) motion is presented here.

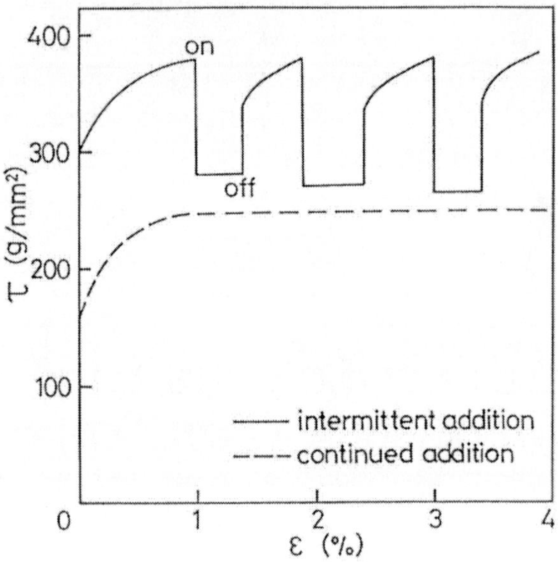

Figure 2-1. Variation of the stress-strain curve for Zn single crystals by superimposed ultrasonic vibrations [32].

The Blaha effect is the phenomenon that static flow stress decreases when an ultrasonic oscillatory stress is superimposed [32]. Blaha and Langenecker found it when the ultrasonic oscillatory stress of 800 kHz was superimposed during plastic deformation of Zn single crystals. Figure 2-1 shows the variation of the stress-strain curve by superimposed ultrasonic vibrations as an example of their experimental results [20]. The curve represents the intermittent addition of ultrasonic vibrations in a solid line and the continued addition in a broken line. This phenomenon is so called the Blaha effect. The same phenomena have been confirmed in many metals [33-35] and have been widely made to apply to the plastic working technique for industrial purpose such as wire drawing, deep drawing, rolling, and another metal forming techniques, since this phenomenon has an industrial significance [36-45]. The Blaha effect has been explained by a temperature rise of materials [39, 46], an abrupt increase in mobile dislocation density [47], a reduction in internal stress [48-50], or a stress superposition mechanism [51-56] so far. The temperature rise is not so high as to influence the flow stress in our tests. It is also improbable that the mobile dislocation density markedly increases immediately after the application of oscillations. If the Blaha effect is explained by itself, the mobile dislocation density has to increase by a factor of tens or hundreds. The reduction in internal stress should be valid only for the particular case that the amplitude of oscillatory stress is so large as to change the internal structure, where a redistribution of the dislocations may occur. The stress superposition mechanism, where the internal friction due to application of oscillations is neglected and the amplitude dependent region (Q_H^{-1}) of internal friction is not considered, has been regarded as the most important one to interpret the Blaha effect. However, Lebedev et al. [50, 57] have reported that the stress decrement ($\Delta\tau$) due to the application of ultrasonic oscillatory stress appears at the moment when measurements of the amplitude dependence of internal friction were made. Ohgaku and Takeuchi [58] also concluded that $\Delta\tau$ is proportional to Q_H^{-1} at small amplitude for NaCl and KCl single crystals. The relation between $\ln\Delta\tau$ and $\ln Q_H^{-1}$ is shown in figure 2-2, which is drawn from the experimental results for single crystals of KCl, NaCl, NaBr, and KBr. The relation is divided into three regions denoted by I, II, and III. It has been furthermore found that the average length of dislocation segments become long when the internal friction measurement was carried out [59]. Therefore, it could be considered that $\Delta\tau$ of alkali halide is attributed to the increase in activation volume for dislocation motion when dislocations move with the aid of thermal activation [58, 60], because the increment in activation volume (υ^*), is caused by the increase in the average length of dislocation segments (L) as given by

$$\Delta\upsilon^* = b\Delta Ld\,, \tag{2-1}$$

where b is the magnitude of Burgers vector and d is the activation distance.

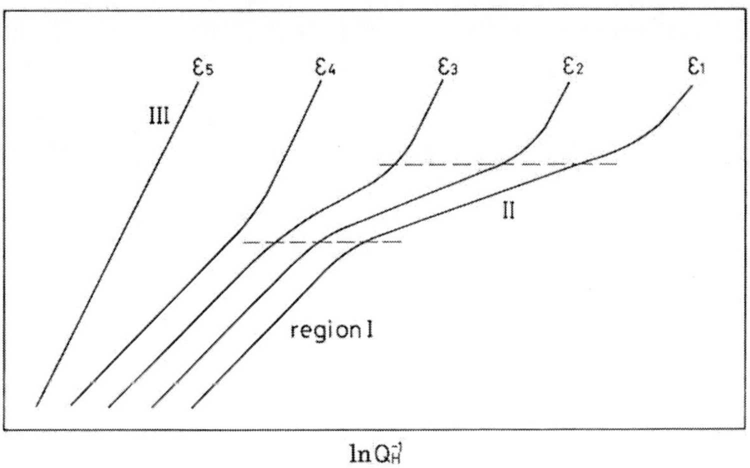

Figure 2-2. Schematic relation between ln$\Delta\tau$ and lnQ_H^{-1} [64]. The strain, ε, increases with its suffix.

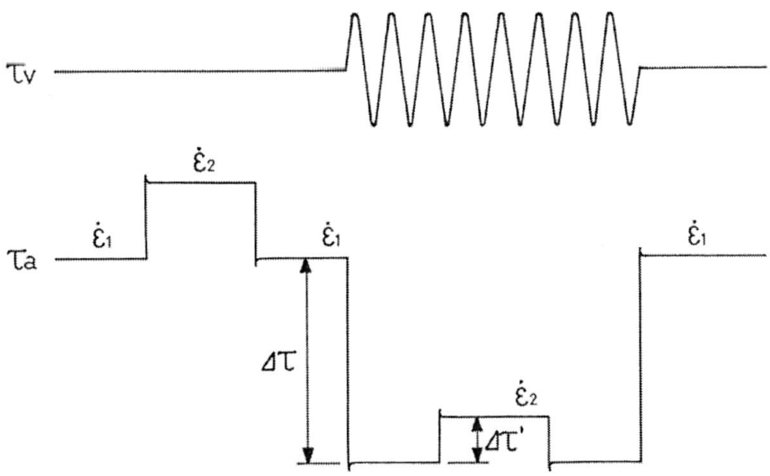

Figure 2-3. Variation of applied shear stress (τ_a) when the strain-rate cycling between the strain rates, $\dot{\varepsilon}_1$ (1.1×10^{-5} s^{-1}) and $\dot{\varepsilon}_2$ (5.5×10^{-5} s^{-1}), is carried out under superposition of ultrasonic oscillatory stress, τ_v.

The $\Delta\tau$ increases with increasing stress amplitude and slightly increases with strain rate ($\dot{\varepsilon}$) [60]. The strain rate dependence of the $\Delta\tau$ may suggest that the stress change ($\Delta\tau'$) due to the strain-rate cycling depend on the stress amplitude and on the $\Delta\tau$. Then, Ohgaku and Takeuchi [60] have carried out the strain-rate cycling tests under superposition of an oscillatory stress.

The strain-rate cycling test during the Blaha effect measurement is illustrated in figure 2-3. The stress drop ($\Delta\tau$) is caused by superposition of oscillatory stress (τ_v) during plastic deformation. When strain-rate cycling between the strain rates of $\dot{\varepsilon}_1$ and $\dot{\varepsilon}_2$ is carried out keeping the stress amplitude of τ_v constant, the stress change due to the strain-rate cycling is $\Delta\tau'$. Figure 2-4 shows the result at room temperature for single crystals of nominally pure alkali halide such as NaCl, KCl, NaBr, and KBr. The relative curve between $\Delta\tau'$ and $\Delta\tau$ shifts upward with strain and bends at a critical $\Delta\tau$. The critical $\Delta\tau$ corresponds to the point which separates region I and region II in figure 2-2. They identified the small $\Delta\tau$ region in figure 2-4 with region I in figure 2-2 and the larger $\Delta\tau$ region with regions II and III. The relations between $\Delta\tau$ and Q_H^{-1} as well as $\Delta\tau$ and $\Delta\tau'$ are essentially divided into two regions, where different mechanisms were considered to dominate. In region III, it concerns the dislocation density, because Q_H^{-1} linearly decreases as applied shear stress, which is generally proportional to the square root of the dislocation density [61-63], becomes large in region III [64]. When the tests are carried out, $\Delta\tau'$ can be expressed by

$$\Delta\tau'=\Delta\tau'_i+\Delta\tau'_p+\Delta\tau'_s, \tag{2-2}$$

where the suffixes of stress variation, i, p, and s, mean the internal stress, the part of the effective stress due to impurities, and the strain-dependent part of the effective stress, respectively. They considered figure 2-4 on the basis of equation (2-2) as follows. $\Delta\tau'_p$ corresponds to the linearly decreasing part of $\Delta\tau'$ at small stress amplitude in figure 2-4 and is given by the difference between $\Delta\tau'$ at $\Delta\tau$ of zero and $\Delta\tau'$ at the critical point which distinguishes regions I and II. $\Delta\tau'_s$ dominates in regions II and III, where $\Delta\tau'_p$ is zero. $\Delta\tau'_s$ increases with increasing strain and dislocation density. With regard to $\Delta\tau'_i$, it was considered by the calculation based on the paper [65] that the influence of internal stress during the strain-rate cycling under the oscillation is negligible. Therefore, the strain-rate cycling tests during the Blaha effect measurement seemed to separate the contributions arising from the interaction between a dislocation and impurities and those from dislocations themselves during plastic deformation.

Ohgaku and Takeuchi [58, 64] have reported that the strain-rate cycling tests associated with the ultrasonic oscillation can separate the contributions arising from the interaction between a dislocation and impurities and from the dislocations themselves during plastic deformation at room temperature. After that, they have discussed the temperature dependence of the effective stress due to monovalent impurities in KCl:Br^- (0.5, 1.0, and 2.0 mol.% in the melt) or I^- (0.2, 0.5, and 1.0 mol.% in the melt) single crystals at 77 to 420 K [66] and NaCl:Br^- (0.1, 0.5, and 1.0 mol.% in the melt) single crystals at 77 to 253 K [67]. They found that the measurement of strain-rate sensitivity under the application of ultrasonic oscillatory stress provides useful information on the interaction between a mobile dislocation and impurities [66, 67].

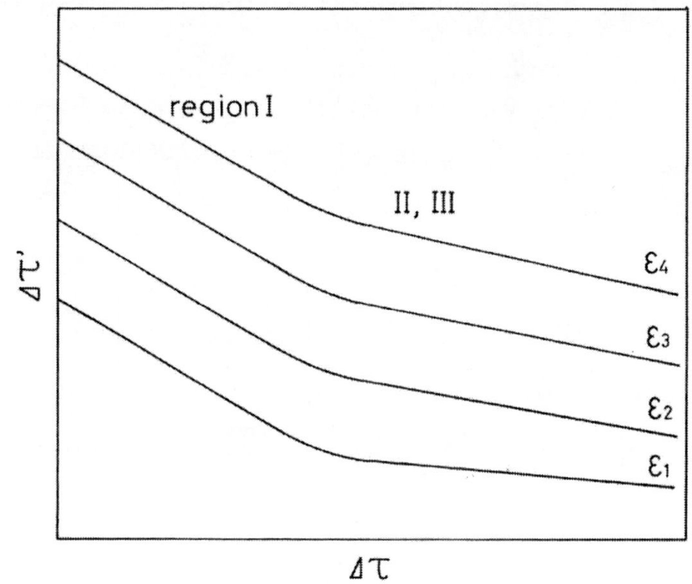

Figure 2-4. Schematic relation between $\Delta\tau'$ and $\Delta\tau$ at room temperature [64]. The strain increases with its suffix.

2.2. Experimental Procedure

A schematic illustration of the apparatus is shown in figure 2-5. A resonator composed of a vibrator and a horn with the resonant frequency of 20 kHz was attached to the testing machine, Shimadzu DSS-500. The specimen, of which the upper and bottom sides are applied with an adhesive agent at room temperature to prevent from slipping during the test, was lightly fixed on a piezoelectric transducer and then cooled down to a test temperature. The specimens were deformed by compression along the <100> axis and the ultrasonic oscillatory stress was intermittently superimposed for one or two minutes by the resonator in the same direction as the compression. The amplitude of the oscillatory stress was observed by a piezoelectric transducer. The strain of the specimen is considered to be homogeneous, because the wavelength, which is 225 mm, is 15 times as long as the length of the specimen.

The strain-rate cycling test during the Blaha effect measurement is illustrated in figure 2-6. Superposition of oscillatory stress (τ_v) during plastic deformation causes a stress drop ($\Delta\tau$). Keeping the stress amplitude of τ_v constant, strain-rate cycling between the strain rates of $\dot{\varepsilon}_1$ (1.1×10^{-5} s^{-1}) and $\dot{\varepsilon}_2$ (5.5×10^{-5} s^{-1}) was carried out. Then, the stress change due to the strain-rate cycling is $\Delta\tau'$. The strain-rate cycling tests made between the crosshead speeds of 20 and 100 μm min^{-1} were performed at 77 K to room temperature. For the tests at 77 K, the specimen was immersed in the liquid nitrogen. The other temperature measurements were made by thermocouples of nickel-55 % copper alloy vs. copper. Each specimen was held at the test temperature for 30 min prior to the compression test and the stability of temperature during the test was kept within 2 K. The strain-rate sensitivity ($\Delta\tau'/\Delta\ln\dot{\varepsilon}$) of the flow stress, which is given by $\Delta\tau'/1.609$, was used as a measure of the strain-rate sensitivity.

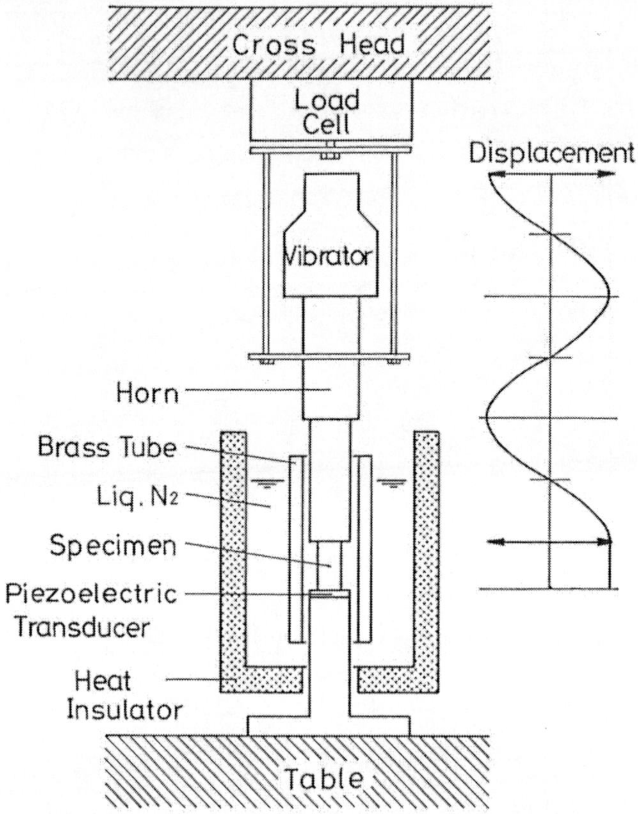

Figure 2-5. Schematic illustration of the apparatus.

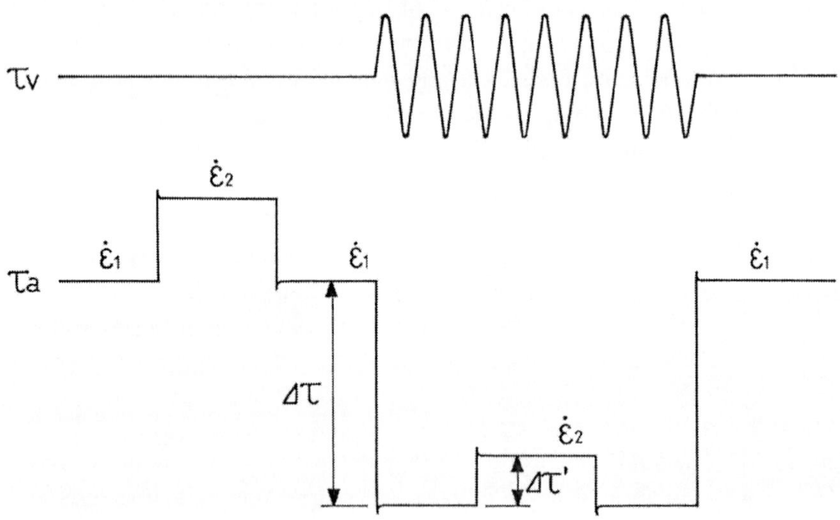

Figure 2-6. Variation of applied shear stress (τ_a) when the strain-rate cycling between the strain rates, $\dot{\varepsilon}_1$ (1.1×10^{-5} s^{-1}) and $\dot{\varepsilon}_2$ (5.5×10^{-5} s^{-1}), is carried out under superposition of ultrasonic oscillatory stress, τ_v.

2.3. Preparation of Specimens

Three kinds of single crystals used in this chapter, KCl, KCl doped with Sr^{2+} (0.035, 0.050, 0.065 mol.% in the melt), and KCl doped with various impurities: sodium, calcium, manganese, nickel, strontium, silver, cesium, barium, thallium and lead (0.050 mol.%, respectively, in the melt), were grown from the melt of a superfine reagent of powders by the Kyropoulos method in air. The specimens, which were obtained from the ingots by cleaving to the size 5×5×15 mm^3, were annealed at 973 K for 24 h in order to reduce dislocation density as far as possible, followed by cooling to room temperature at the rate of 40 Kh^{-1}. Furthermore, the specimens were held at 673 K for 30 min, followed by quenching to room temperature immediately before the test, in order to disperse the impurities into them. The temperature of this heat treatment is determined for the below reason.

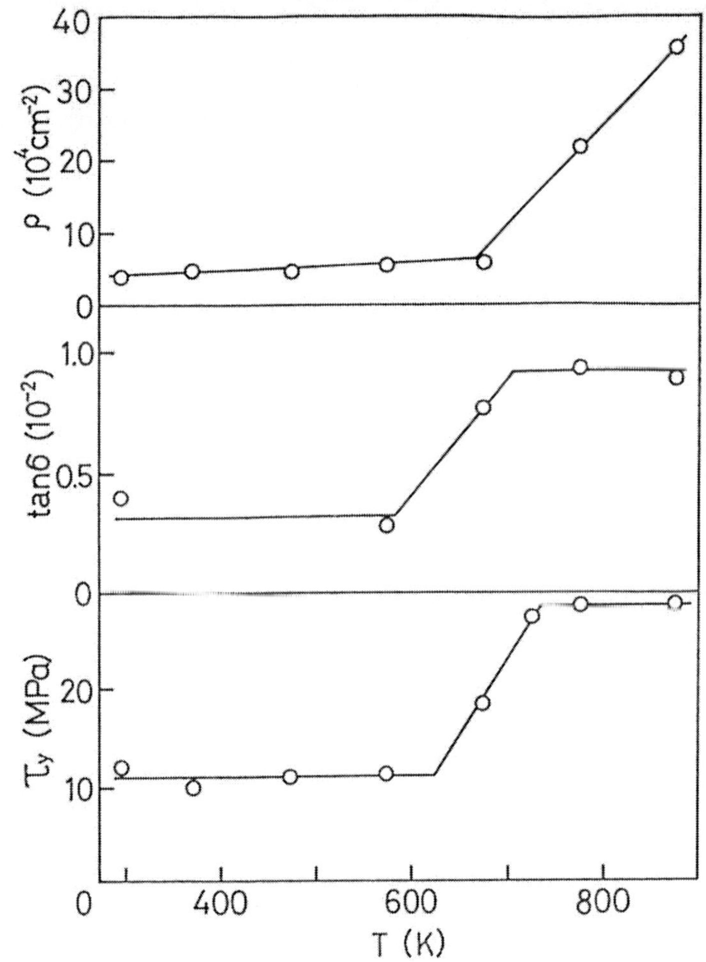

Figure 2-7. Dependence of dislocation density (ρ), tan δ , and yield stress (τ_y) on the annealing temperature [68].

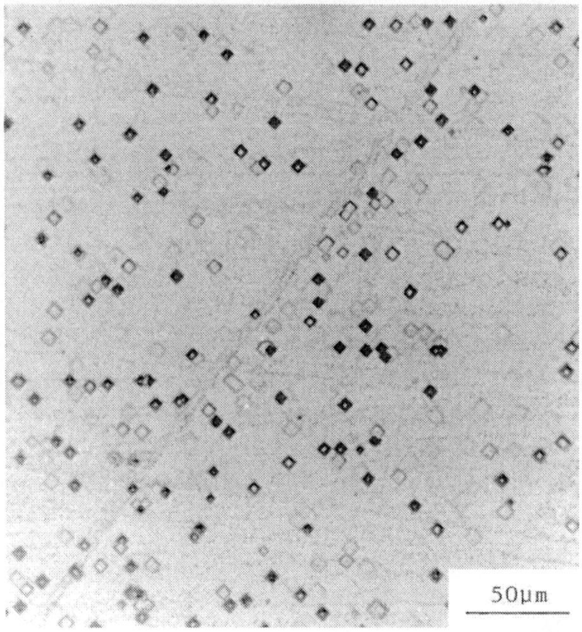

Figure 2-8. The optical micrograph of the dislocations on a (100) plane for the annealed KCl:Sr^{2+} (0.050 mol.%) before the tests.

Katoh [68] has examined the dependence of dislocation density (ρ), tanδ, and yield stress (τ_y) on the quenching temperature for KCl:Sr^{2+}(0.3 and 1.0 mol.% in the melt) single crystals within 373 to 873 K. The result is shown in figure 2-7. The measurement of dislocation density in a crystal has been carried out by the etch pit technique. Acorrosive liquid was a saturated solution of $PbCl_2$+ethyl alcohol added two drops of water. The etching was carried out at room temperature for 30 min. However, it is difficult to resolve the individual etch pits for high dislocation density. Figure 2-8 shows the optical micrograph of the etch pits for KCl:Sr^{2+} (0.050 mol.% in the melt) annealed at 973 K for 24 h. The position of the dislocation after the annealed treatment is marked by a pyramidal pit, and the position where a dislocation slipped out of the crystal after the treatment is marked by a flat-bottom pit. The dislocation density on a (100) plane is found to be 1.27×10^4 cm^{-2} for the annealed specimen. X-ray topography has also been used, but there is lack of resolution in the photograph for large deformation. Therefore, the specimens for these methods were limited to low dislocation density. Although the electron microscopy could be used to detect dislocations, it has provided the information for a thin specimen but not for bulk.

As a result of figure 2-7, it was found that the dislocation density and the yield stress remarkably increase for the crystals quenched from the temperature above 673 K and the value of tanδ does not vary by quenching from the temperature below 573 K or above 673 K. The concentration of isolated I-V dipoles affects the yield stress, as reported in the papers [1, 2, 8, 69-73]. The difference in dislocation density is slight and the tanδ obviously becomes larger with higher quenching temperature between 573 and 673 K. Therefore, the specimens are determined to be quenched from 673 K to room temperature immediately before the compression tests.

2.4. $\Delta\tau'/\Delta \ln\dot{\varepsilon}$ Versus $\Delta\tau$ for Various Specimens

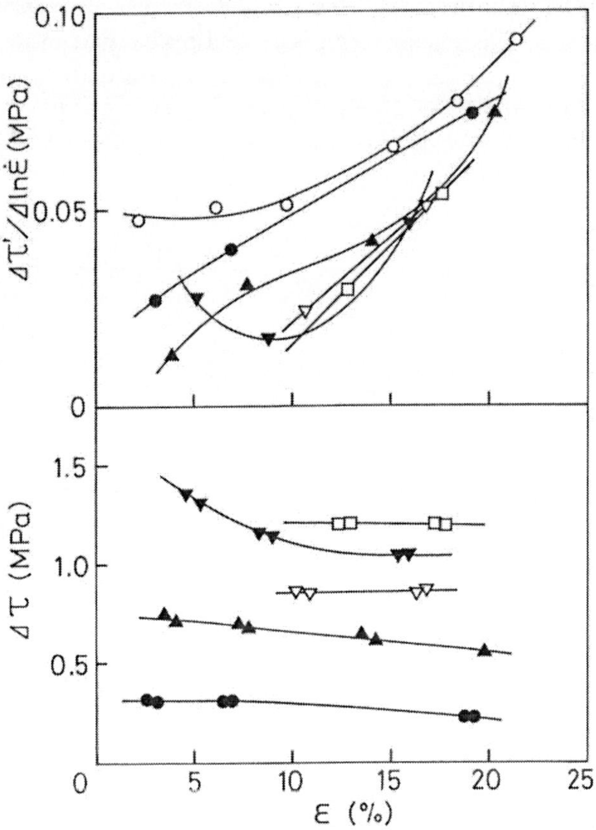

Figure 2-9. Variation of the strain-rate sensitivity, $\Delta\tau'/\Delta\ln\dot{\varepsilon}$, and the stress decrement, $\Delta\tau$, with shear strain for KCl:Sr^{2+} (0.050 mol.%) at 200 K. τ_v (arb. units): (○) 0, (●) 10, (▲) 25, (▽) 35, (▼) 45, and (□) 50.

Figure 2-9 shows the variation of the strain-rate sensitivity and the stress decrement with the shear strain for KCl:Sr^{2+} (0.050 mol.%) single crystals at 200 K. $\Delta\tau$ does not change significantly with shear strain but increases with stress amplitude at a given temperature and shear strain. $\Delta\tau'/\Delta\ln\dot{\varepsilon}$ tends to increase with shear strain and decrease with stress amplitude. The increase of strain-rate sensitivity with shear strain is caused by the increase of the forest dislocation density. The relation between the strain-rate sensitivity and the stress decrement obtained from figure 2-9 at a strain of 10, 14 and 18 % is shown in figure 2-10, which provides the relationship between $\Delta\tau'/\Delta\ln\dot{\varepsilon}$ and $\Delta\tau$ for a fixed internal structure of the KCl:Sr^{2+} (0.050 mol.%) crystal at 200 K. Figure 2-11 concerns several temperatures, which is obtained by the above-mentioned method. As can be seen from figures 2-10 and 2-11, the variation of strain-rate sensitivity with stress decrement is stair-like (there are two bending points on each curve, and there are two plateau regions: the first plateau region ranges below the first bending point (τ_{p1}) at low stress decrement and the second one extends from the second bending point (τ_{p2}) at high stress decrement). The second plateau region is considered to correspond to the regions II and III in figure 2-4, as reported by Ohgaku and Takeuchi [74].

The strain-rate sensitivity decreases with the stress decrement between the two bending points. Furthermore, figure 2-10 shows the influence of the shear strain on the relation between the strain-rate sensitivity and the stress decrement. The curve shifts upwards with increasing shear strain. This phenomenon is caused by the part of the strain-rate sensitivity which depends on dislocation cuttings. Since the dislocation cuttings increase with increasing strain, the strain-rate sensitivity increases [74]. Figure 2-11 shows the influence of temperature on the relation between the strain-rate sensitivity and the stress decrement. As the temperature is lower, τ_{p1} and τ_{p2} are larger.

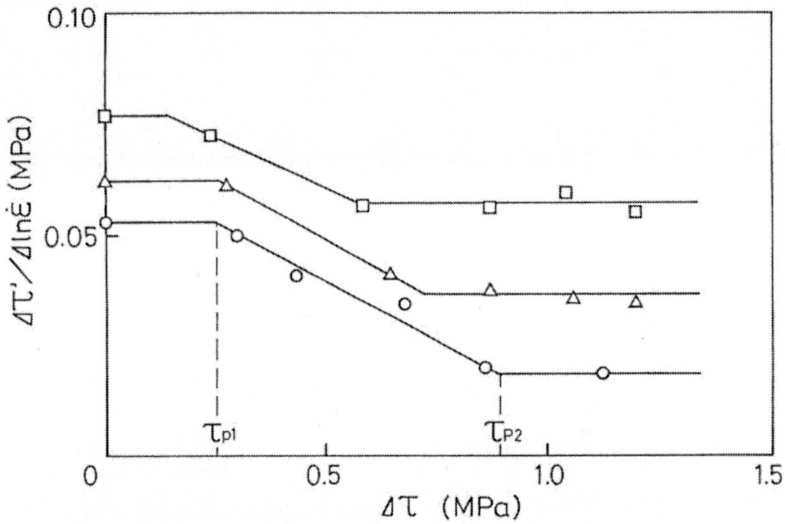

Figure 2-10. Relation between the strain-rate sensitivity and the stress decrement for KCl:Sr^{2+} (0.050 mol.%) at 200 K. ε : (○) 10%, (Δ) 14%, and (□) 18%.

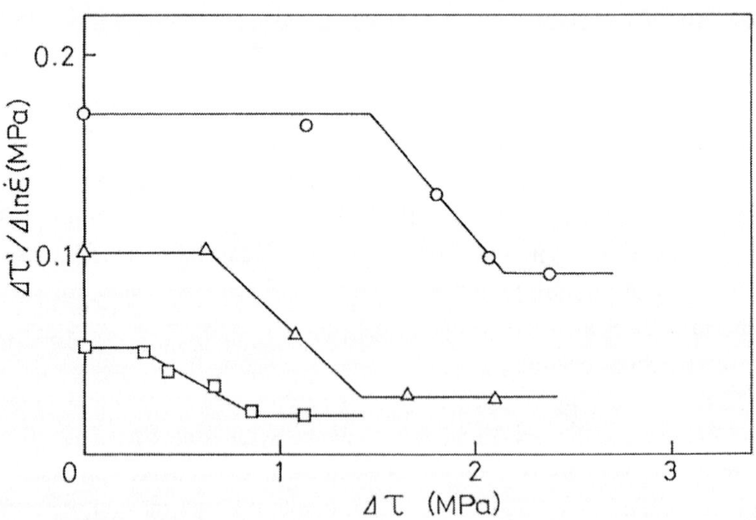

Figure 2-11. Relation between the strain-rate sensitivity and the stress decrement for KCl:Sr^{2+} (0.050 mol.%) at various temperatures: (○) 103 K, ε =9 %; (Δ) 133 K, ε =8 %; (□) 200 K, ε =10 %.

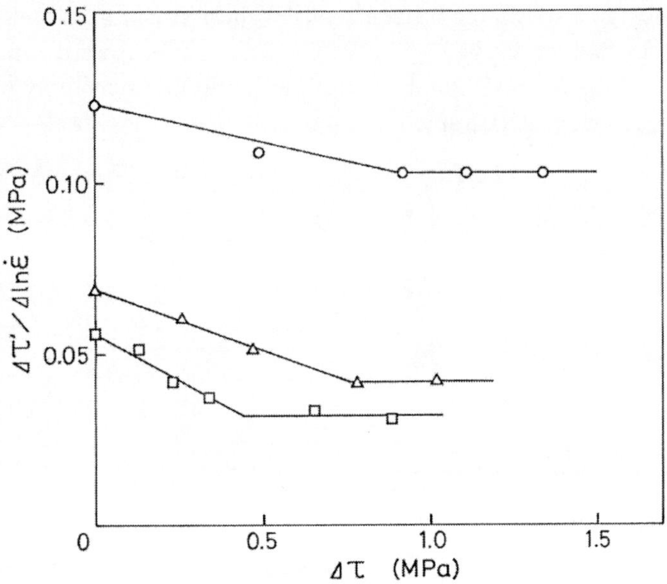

Figure 2-12. Relation between the strain-rate sensitivity and the stress decrement for KCl at various temperatures: (○) 77 K, (Δ) 94 K, (□) 168 K.

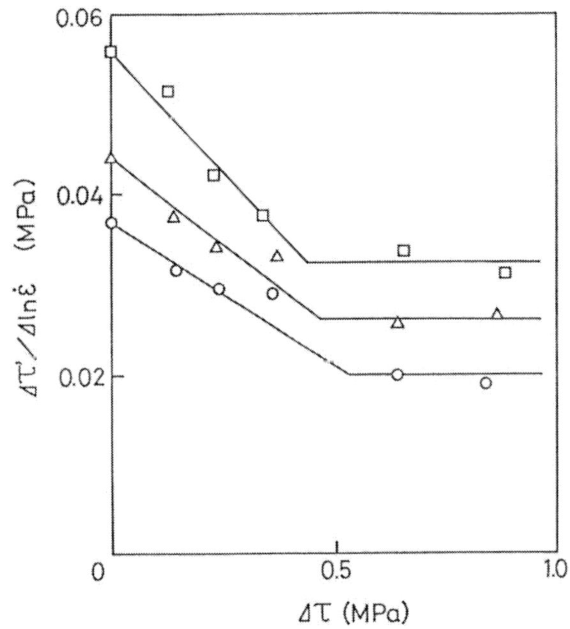

Figure 2-13. Relation between the strain-rate sensitivity and the stress decrement for KCl at 168 K. ε : (○) 10 %, (Δ) 12 %, and (□) 15 %.

Figure 2-12 shows the relation between the strain-rate sensitivity and the stress decrement for nominally pure KCl single crystals at several temperatures. In contrast to $KCl:Sr^{2+}$ single crystals, there is only one bending point on each curve, which is considered to correspond to τ_{p2}. The bending point shifts in the direction of higher stress decrement with decreasing temperature. Figure 2-13 shows the influence of the shear strain on the relation between the strain-rate sensitivity and the stress decrement. The curve shifts upward with increasing shear strain. This phenomenon is caused by the part of the strain-rate sensitivity which depends on dislocation cuttings as well as $KCl:Sr^{2+}$.

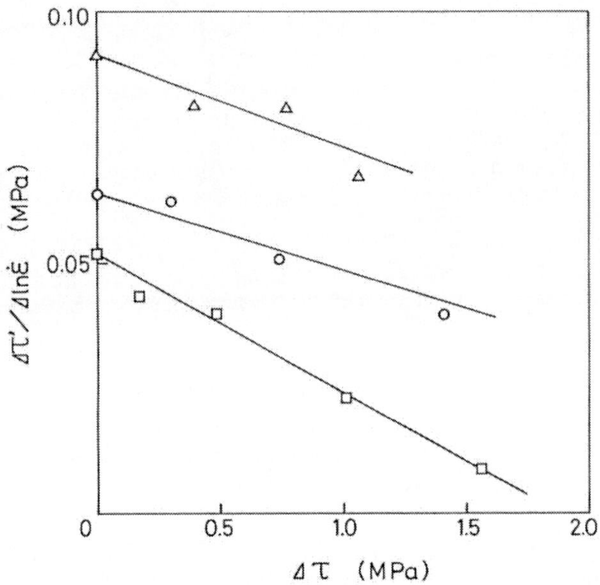

Figure 2-14. Relation between the strain-rate sensitivity and the stress decrement for KCl doped with various impurities at various temperatures: (○) 151 K, (Δ) 159 K, (□) 195 K.

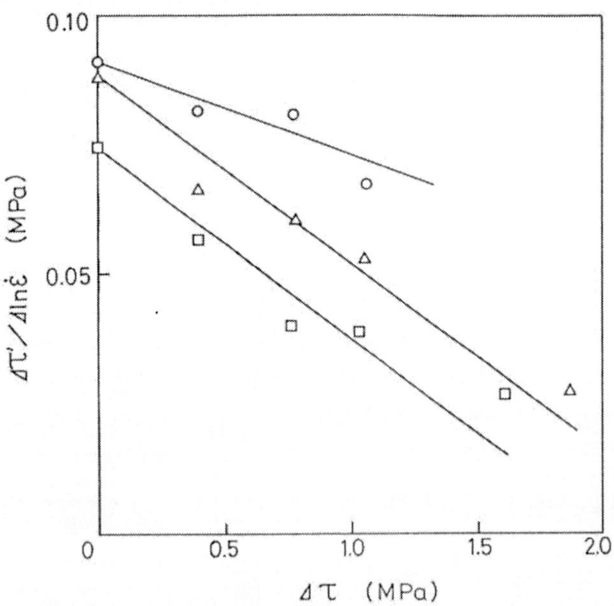

Figure 2-15. Relation between the strain-rate sensitivity and the stress decrement for KCl doped with various impurities at 159 K. ε : (○) 8 %, (Δ) 10 %, and (□) 12 %.

Figures 2-14 and 2-15 show the relation between the strain-rate sensitivity and the stress decrement for KCl single crystals containing various impurities. Figures 2-14 and 2-15 concern several temperatures and shear strains, respectively. It is shown in these figures that no bending point appears for the specimens. The strain-rate sensitivity only decreases with increasing stress decrement.

2.5. τ_{P1} AND τ_{P2} VERSUS TEMPERATURE FOR KCL:SR^{2+} SINGLE CRYSTALS

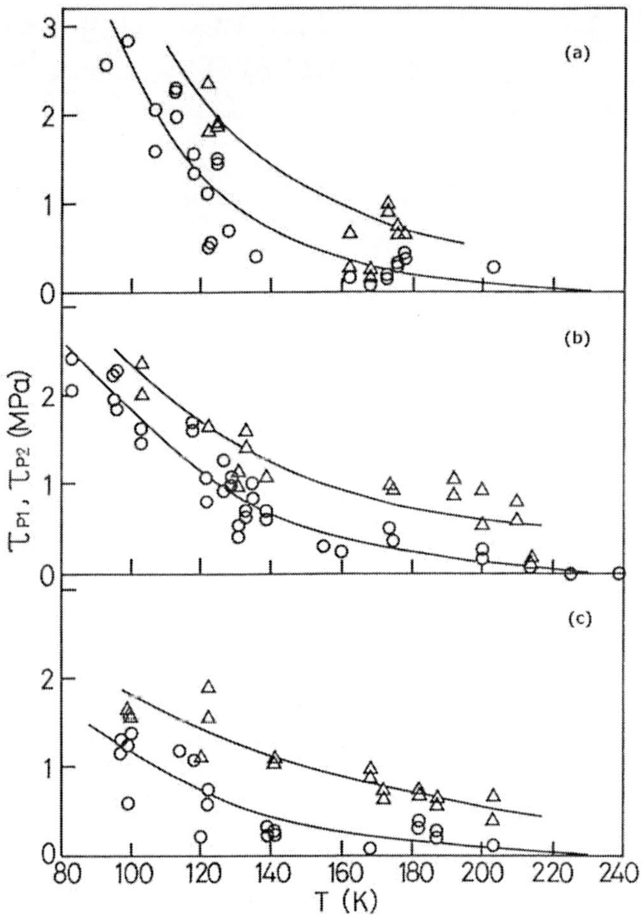

Figure 2-16. Dependence of (○) τ_{p1} and (Δ) τ_{p2} on temperature for KCl:Sr^{2+} single crystals. The Sr^{2+} concentration is (a) 0.065, (b) 0.050, and (c) 0.035 mol.%.

If τ_{p1} is the effective stress due to the weak obstacles (i.e. divalent impurities) on the mobile dislocation during plastic deformation of KCl:Sr^{2+} single crystals, it should depend on temperature and impurity concentration. Figures 2-16 (a), (b) and (c) show the dependence of τ_{p1} and τ_{p2} on temperature for KCl:Sr^{2+} (0.065, 0.050 and 0.035 mol.%, respectively) single crystals. It is clear from the figures that τ_{p1} and τ_{p2} tend to increase with decreasing temperature for three specimens. Both τ_{p1} and τ_{p2} increase with increasing concentration of

Sr^{2+} at a given temperature. The critical temperature (T_c), at which the curves intersect the abscissa and τ_{p1} is zero, may be determined from these figures. Then, T_c appears to be constant independently of the Sr^{2+} concentration. It is clear from these phenomena that τ_{p1} corresponds to the effective stress due to the weak obstacles which lie on the dislocation when the dislocation moves forward with the help of oscillation. Observation of τ_{p1}, therefore, provides information on the interaction between a dislocation and weak obstacles. That is to say, the temperature dependence of τ_{p1} reveals the force-distance profile which expresses the interaction between a dislocation and weak obstacle.

2.6. τ_{P1} And τ_{P2} Versus Yield Stress for KCl:Sr^{2+} Single Crystals

It is clear from figures 2-10 to 2-15 that the curve of the strain-rate sensitivity and the stress decrement has two bending points and two plateau regions only for KCl:Sr^{2+}, which is considered to have only one type of impurity. Thus, τ_{p1} and τ_{p2} may depend on impurity concentration. Figure 2-17 shows the dependence of τ_{p1} and τ_{p2} on the yield stress for KCl:Sr^{2+} at 150 K. The plots correspond to the case when the Sr^{2+} concentration is 0.035, 0.050, and 0.065 mol.% from the bottom. It can be seen from this figure that both τ_{p1} and τ_{p2} are approximately proportional to the yield stress. This means that τ_{p1} and τ_{p2} increase, depending on the impurity concentration, because the yield stress generally increases with increasing impurity concentration [1-5], as can be seen from the dependence of the yield stress on temperature in figure 2-18.

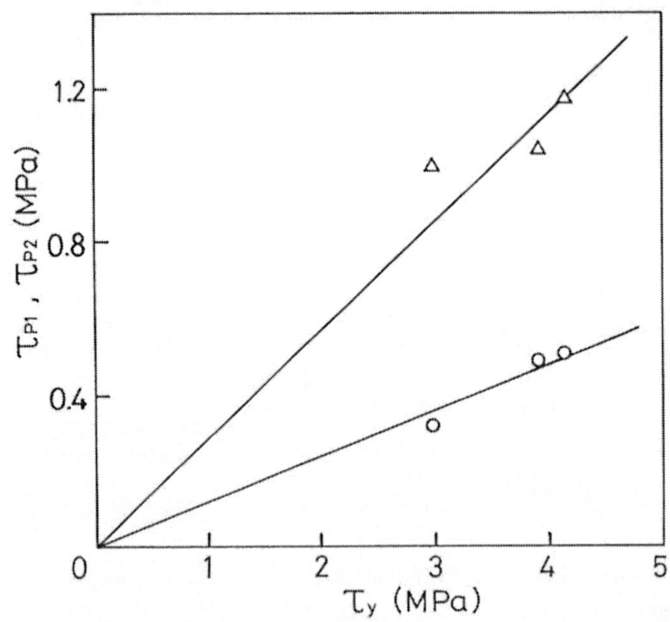

Figure 2-17. Dependence of (○) τ_{p1} and (Δ) τ_{p2} on the yield stress (τ_y) for KCl:Sr^{2+} at 150 K. Concentration of Sr^{2+} is 0.035, 0.050, 0.065 mol.% from the left.

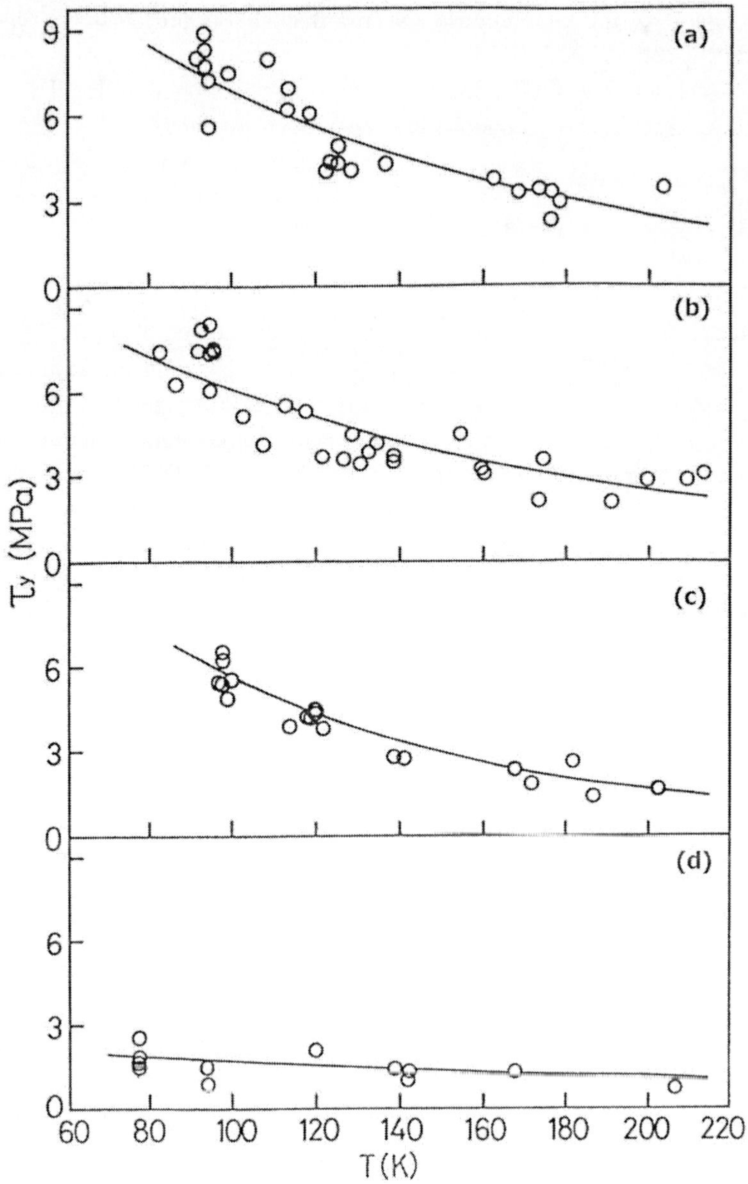

Figure 2-18. Dependence of the yield stress (τ_y) on temperature for KCl:Sr^{2+} ((a) 0.065, (b) 0.050, (c) 0.035 mol.% in the melt) and (d) KCl.

2.7. Discussion for the Relation between the Strain-Rate Sensitivity and the Stress Decrement for KCl:Sr^{2+} Single Crystals

The above experimental facts that the curve of the strain-rate sensitivity and stress decrement for only KCl:Sr^{2+} has two bending points and two plateau regions, and both τ_{p1} and τ_{p2} depend on Sr^{2+} concentration, suggest that the phenomena shown in figures 2-10 and

2-11 are attributable to the interaction between a dislocation and only one type of obstacle. The reason for this is discussed below.

The strain-rate sensitivity of effective stress is inversely proportional to the activation volume, i.e. the average length of dislocation segment, as given by

$$\left(\frac{\Delta\tau^*}{\Delta \ln \dot{\varepsilon}}\right)_T = \frac{kT}{\upsilon^*}, \ (\upsilon^* = bLd) \tag{2-3}$$

where k is the Boltzmann constant and T is an absolute temperature. In addition, it is reported that the length of the dislocation segment increases and the strain-rate sensitivity decreases when the ultrasonic oscillatory stress is applied at room temperature during plastic deformation and that the plateau region is due to the dislocation cuttings when oscillations cause the dislocation to break away from all weak obstacles [74]. Therefore, the first plateau region, as well as the second one, indicates that the average length of the dislocation segments remains constant there. Thus, application of oscillations with low stress amplitude cannot influence the average length of the dislocation segments at low temperature, but it can do so at high temperature, such as room temperature. Therefore, the plateau region appears at low stress decrement in figures 2-10 and 2-11. Now it is imaged that a dislocation is pinned by many weak obstacles and is bowing by applied stress between a few strong obstacles such as forest dislocations during stationary plastic deformation. When the stress amplitude increases, the dislocation begins to break away from weak obstacles by oscillation between the strong ones and the average length of the dislocation begins to increase. The strain-rate sensitivity starts to decrease at the stress decrement of τ_{p1}. This τ_{p1} should depend on temperature, and on type and density of the obstacle. Consequently, the phenomena shown in figures 2-10 to 2-15 reflect the influence of ultrasonic oscillation on the dislocation motion on the slip plane containing many weak obstacles and a few strong ones, during plastic deformation. Furthermore, τ_{p1} is considered to represent the effective stress due to the weak obstacles which lie on the dislocation when the dislocation begins to break away from these weak obstacles with the help of oscillation. The weak obstacles are considered to be the impurities and not to be vacancies here, because the vacancies in the specimen have low density in contrast to the impurities. For example, the concentration of vacancies at thermal equilibrium state is about 1.2×10^{-7} % for KCl single crystal quenched from 673 K to room temperature. As the stress amplitude increases further, oscillating dislocations overcome all Sr^{2+} ions (i.e., Sr^{2+} ion does not act as an obstacle to the translational motion of dislocations) and the strain-rate sensitivity becomes constant, as reported for the plastic deformation of RbCl doped with Br^- or I^- by Ohgaku and Hashimoto [75]. The stress decrement at which the ultrasonic oscillatory stress helps the dislocation break away from all weak obstacles is τ_{p2}. At a stress decrement more than τ_{p2}, forest dislocations remain as the obstacles to the mobile dislocation.

The curve of strain-rate sensitivity and the stress decrement for the specimen containing many types of obstacle should be obtained by superimposition of various curves for each obstacle. Therefore, the bending points and plateaus should not be clear. The lines shown in figures 2-14 and 2-15 correspond to this case. KCl crystal (nominal pure KCl crystal) may contain a small amount of various impurities, although none were added. As a result, the first bending point does not appear. The second bending point, however, appears because the amount of impurities is small.

CONCLUSION

When the strain-rate cycling test during the Blaha effect measurement is carried out for KCl:Sr^{2+}, the dependence of the strain-rate sensitivity on the stress decrement has two bending points and two plateau regions. The plots of the strain-rate sensitivity and the stress decrement due to oscillation, which can be obtained through the combination method, seems to reflect the influence of ultrasonic oscillation on the dislocation motion on the slip plane containing many impurities and a few forest dislocations, and further seems to separate the contributions arising from the interaction between a dislocation and impurities and those from dislocations themselves during plastic deformation of KCl:Sr^{2+}. The first bending point, τ_{p1}, at low stress decrement depends on type and density of the weak obstacles and corresponds to the effective stress due to weak obstacles on the mobile dislocation.

REFERENCES

[1] Suszyńska, M. *Kristall. Technik.* 1974, *9*, 1199.

[2] Dryden, J. S.; Morimoto, S.; Cook, J. S. *Phil. Mag.* 1965, *12*, 379.

[3] Stoloff, N. S.; Lezius, D. K.; Johnston, T. L. *J. Appl. Phys.* 1963, *34*, 3315.

[4] Chin, G. Y.; Van Uitert, L. G.; Green, M. L.; Zydzik, G. J.; Kometani, T. Y. *J. Amer. Ceram. Soc.* 1973, *56*, 369.

[5] Kataoka, T.; Yamada, T. *Jpn. J. Appl. Phys.* 1977, *16*, 1119.

[6] Boyarskaya, Yu. S.; Zhitaru, R. P.; Palistrant, N. A. *Cryst. Res. Technol.* 1990, *25*, 1469.

[7] Boyarskaya, Yu. S.; Zhitaru, R. P.; Palistrant, N. A. *Sov. Phys. Solid State* 1990, *32*, 1989.

[8] Pratt, P. L.; Harrison, R. P.; Newey, C. W. A. *Disc. Faraday Soc.* 1964, *38*, 211.

[9] Newey, C. W. A.; Harrison, R. P.; Pratt, P. L. *Proc. Brit. Ceram. Soc.* 1966, *6,* 305.

[10] Chin, G. Y.; Van Uitert, L. G.; Green, M. L.; Zydzik, G. *Scripta Metall.* 1972, *6,* 475.

[11] Green, M. L.; Zydzik, G. *Scripta Metall.* 1972, *6*, 991.

[12] Andreev, G. A.; Klimov, V. A. *Sov. Phys. Solid State* 1980, *22*, 2042.

[13] Buravleva, M. G.; Rozenberg, G. Kh.; Soifer, L. M.; Chaikovskii, E. F. *Sov. Phys. Solid State* 1980, *22*, 150.

[14] Narasimha Reddy, K.; Subba Rao, U. V. *Cryst. Res. Technol.* 1984, *19*, K73.

[15] Appel, F.; Messerschmidt, U.; Nadgorny, E. M.; Zaitsev, S. I. *Mater. Sci. Engng.* 1982, *52*, 69.

[16] Strunk, H. *Phys. Status Solidi* (a) 1975, *28,* 119.

[17] Messerschmidt, U.; Appel, F. *Phil. Mag. A* 1985, *51,* 781.

[18] Kataoka, T.; Ohji, H.; Morishita, H.; Kishida, K.; Azuma, K.; Yamada, T. *Jpn. J. Appl. Phys.* 1989, *28*, L697.

[19] Kataoka, T.; Ohji, H.; Kishida, K.; Azuma, K.; Yamada, T. *Appl. Phys. Lett.* 1990, 56 1317.

[20] Kataoka, T. *Butsuri* 1992, *47*, 713 (in Japanese).

[21] Young, F. W. JR. In *Dislocation Dynamics*, (edited by Rosenfield, A. R.; Hahn, G. T.; Bement, A. L. Jr.; Jaffee, R. I.) McGraw-Hill Book Co.; New York, 1968; p 313.

[22] Philips, D. C.; Pratt, P. L. *Phil. Mag.* 1970, *21*, 217.
[23] Philips, D. C.; Pratt, P. L. *Phil. Mag.* 1970, *22,* 809.
[24] Ivanov, V. I.; Lebedev, A. B.; Kardashev, B. K.; Nikanorov, S. P. *Sov. Phys. Solid State* 1986, *28*, 867.
[25] Indenbom, V. L.; Chernov, V. M. *Phys. Status Solidi* (a) 1972, *14,* 347.
[26] Schwarz, R. B.; Granato, A. V. *Phys. Rev. Lett.* 1975, *34,* 1174.
[27] Kosugi, T.; Kino, T. *Jpn. J. Appl. Phys. Suppl.* 1987, *26*, 881.
[28] Dotsenko, V. I. *Phys. Status Solidi* (b) 1979, *93,* 11.
[29] Gupta, I.; Li, J. C. M. *Mater. Sci. Engng.* 1970, *6*, 20.
[30] Johnston, W. G.; Gilman, J. J. *J. Appl. Phys.* 1959, *30*, 129.
[31] Granato, A. V.; Lücke, K. *J. Appl. Phys.* 1956, *27,* 583.
[32] Blaha, F.; Langenecker, B. *Z. Naturwiss.* 1955, *20*, 556.
[33] Langenecker, B. *IEEE Trans. Sonic and Ultrasonic* 1966, SU-*13*, 1.
[34] Nevill, G. E.; Brotzen, F. R. *Proc. ASTM* 1957, *57,* 571.
[35] Izumi, O.; Oyama, K.; Suzuki, Y. *J. Japan Soc. for Tech. of Plasticity* 1966, *7*, 59.
[36] Yamamoto, M.; Fujimori, S.; *Kikai no Kenkyu* 1970, *22*, 1129.
[37] Yamamoto, M.; Fujimori, S.; *Kikai no Kenkyu* 1970, *22*, 1229.
[38] Mori, E. *J. Japan Soc. for Tech. of Plasticity* 1970, *11*, 257.
[39] Yamamoto, M. *J. Japan Soc. for Tech. of Plasticity* 1970, *11*, 239.
[40] Evans, A. E.; Smith, A. W.; Waterhouse, W. J.; Sansome, D. H. *Ultrasonics* 1975, *13,* 162.
[41] Weiss, B. *Aluminium* 1972, *48*, 741.
[42] Weiss, B. *Aluminium* 1972, *48*, 811.
[43] Weiss, B. *Aluminium* 1973, *49*, 165.
[44] Mori, E. *J. Inst. Electronics & Communication Engineers of Japan* 1975, *58*, 781.
[45] Langenecker, B. *Maschinenmarkt* 1975, *81*, 781.
[46] Puskar, A. *Metallurg. Trans.* 1976, *7A,* 1529.
[47] Jon, M. C.; Mason, W. P.; Beshers, D. N. *J. Appl. Phys.* 1976, *47*, 2337.
[48] Shetty, D. K.; Meshii, M. *Metallurg. Trans.* 1975, *6A*, 349.
[49] Brandley, W. A.; Nam, S. W.; Matlock, D. K. *Metallurg. Trans.* 1976, *7A*, 425.
[50] Lebedev, A. B.; Kustov, S. B.; Kardashev, B. K. *Sov. Phys. Solid State* 1982, *24,* 1798.
[51] Baker, G. S.; Carpenter, S. H. *J. Appl. Phys.* 1967, *38,* 1586.
[52] Kaiser, G.; Pechhold, W. *Acta Metall.* 1967, *17*, 527.
[53] Suzuki, T. *Seisan-Kenkyu* 1970, *22*, 194 (in Japanese).
[54] Suzuki, T. *Seisan-Kenkyu* 1970, *22*, 344 (in Japanese).
[55] Endo, T.; Suzuki, K.; Ishikawa, M. *Trans. Japan. Inst. Metals* 1979, *20,* 706.
[56] Endo, T.; Tasaki, M.; Kubo, M.; Shimada, T. *J. Japan. Inst. Metals* 1982, *46*, 773 (in Japanese).
[57] Lebedev, A. B.; Kustov, S. B.; Kardashev, B. K. *Sov. Phys. Solid State* 1983, *25*, 511.
[58] Ohgaku, T.; Takeuchi, N. *Phys. Status Solidi* (a) 1988, *105*, 153.
[59] Hikata, A.; Johnson, R. A.; Elbaum, C. *Phys. Rev.* 1970, *B 2*, 4856.
[60] Ohgaku, T.; Takeuchi, N. *Phys. Status Solidi* (a) 1987, *102*, 293.
[61] Hesse, J.; Hobbs, L. W. *Phys. Status Solidi* 1972, *14*, 599.
[62] Frank, W.; *Mater. Sci. Eng.* 1970, *6*, 121.
[63] Frank, W.; *Mater. Sci. Eng.* 1970, *6*, 132.
[64] Ohgaku, T.; Takeuchi, N. *Phys. Status Solidi* (a) 1989, *111*, 165.

[65] Li, J. C. M. In *Dislocation Dynamics*, (edited by Rosenfield, A. R.; Hahn, G. T.; Bemont, A. L.; Jaffee, R. I.) Mc Graw-Hill Publ. Co.; New York, 1968; p 87.
[66] Ohgaku, T.; Takeuchi, N. *Phys. Status Solidi* (a) 1992, *134*, 397.
[67] Ohgaku, T.; Teraji, H. *Phys. Status Solidi* (a) 2001, *187*, 407.
[68] Katoh, S. *Graduation Thesis at Kanazawa Univ.;* Kanazawa, 1987, pp 16-21 (in Japanese).
[69] Lubenets, S. V.; Startsev, V. I. *Soviet Phys. Solid State*, 1968, *10*, 15.
[70] Orozco, E.; Mendoza, M. A.; Soull, J.; Rubio, J. *Jpn. J. Appl. Phys.* 1982, *21*, 249.
[71] Gilman, J. J. *J. Appl. Phys.* 1974, *45*, 508.
[72] Reddy, B. K. *Phys. Status Solidi* (a) 1987, *99*, K7.
[73] Zaldo, C.; Sole, J. G.; Agullo-Lopez, F. *J. Mater. Sci.* 1982, *17*, 1465.
[74] Ohgaku, T.; Takeuchi, N. *Phys. Status Solidi* (a) 1990, *118*, 153.
[75] Ohgaku, T.; Hashimoto, K. *Mater. Sci. Engng. A* 2005, *400-401*, 401.

Chapter 3

Influence of State of Impurities on the Effective Stress Due to the Impurities

3.1. Introduction

It is well known that the state of impurities in a crystal affect its hardness. Green and Zydzik [1] reported that when a microhardness tester was used with a Knoop diamond indentor, the greatest hardness occurs for the highest concentration of isolated dipole and the decrease of hardness is then attributed to precipitation in $KCl:Sr^{2+}$ single crystals. Dryden et al. [2] also reported that when a small amount of impurities diffuse into the matrix, or they aggregate, their state strongly influences the resistance to movement of the dislocations in various alkali halide crystals doped with divalent cations (NaCl: Ca^{2+} or Mn^{2+}, KCl: Sr^{2+} or Ba^{2+}, $LiF:Mg^{2+}$).

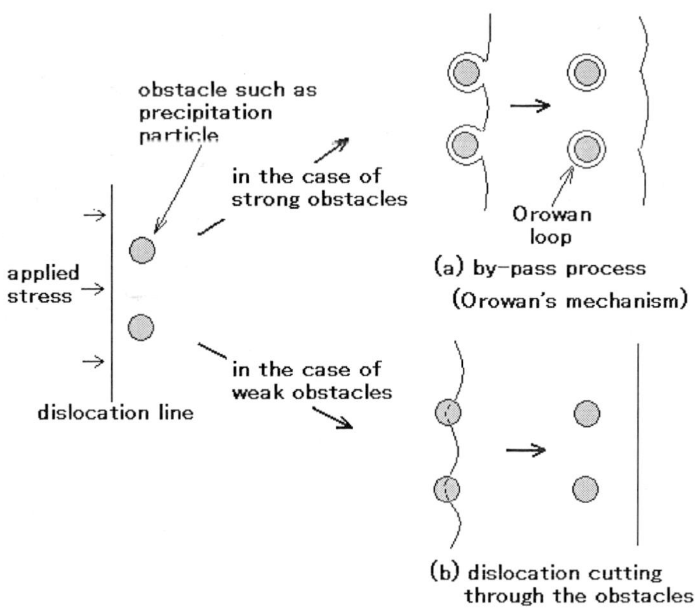

Figure 3-1. Process of dislocation motion on the slip plane containing athermal obstacles: (a) strong obstacles or (b) weak obstacles.

When the dislocation cannot surmount large precipitations even with the aid of thermal fluctuations, Lubenets and Startsev [3] reported that either the dislocation will cut across them or the dislocation will be diffracted at them according to Orowan's mechanism. The process of dislocation motion is schematically illustrated in figure 3-1.

In this chapter, influence of a state of impurities on the interaction between a dislocation and impurities is investigated for $KCl:Sr^{2+}$. This chapter refers to its influence on the effective stress due to the impurities.

3.2. Relation between the Strain-Rate Sensitivity and the Stress Decrement for the Two Kinds of Specimens

Relation between the strain-rate sensitivity and the stress decrement for $KCl:Sr^{2+}$ (0.050 mol.% in the melt) at the shear strain of 10 % is shown in figure 3-2. The measuring temperature is 103 K. Open circles in the figure represent the relation for the quenched specimen used in the chapter 2, and open triangles for the specimen stored at room temperature for a half year. The τ_{p1} for the stored specimen is smaller than that for the quenched specimen. The τ_{p2} for the stored specimen, however, does not appear within the measurement, because high stress amplitude could not be applied to the specimen during the strain-rate cycling tests.

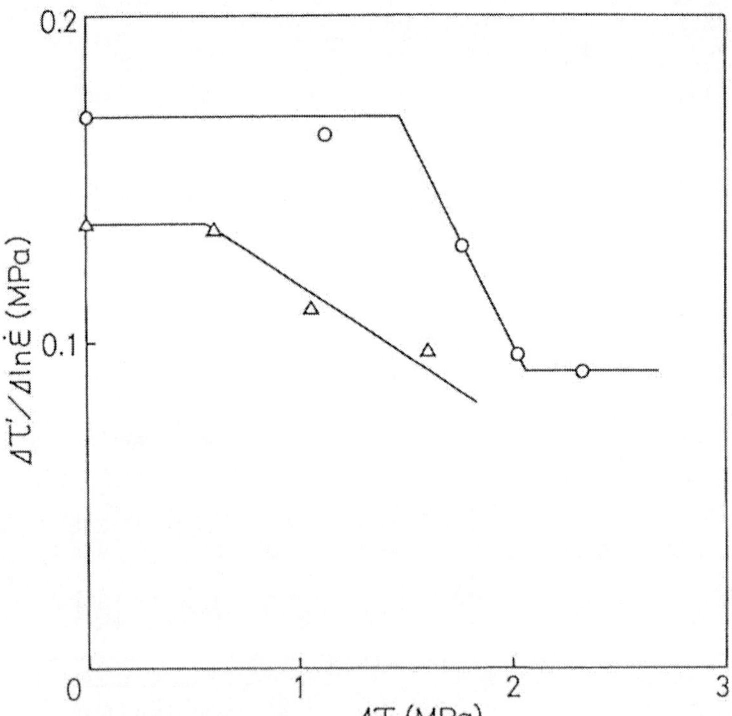

Figure 3-2. Relation between the strain-rate sensitivity and the stress decrement at 103 K for (○) the quenched specimen and (Δ) the stored specimen.

Furthermore, τ_{p1} and τ_{p2} of the quenched and the stored specimens were investigated at various temperatures. Figures 3-3 (a) and (b) show the results for the quenched specimen and for the stored one, respectively. The τ_{p1} for the stored specimen is smaller than that for the quenched specimen at a given temperature. On the other hand, the τ_{p2} for the stored specimen is a little smaller than that for the quenched one as compared with τ_{p1}. In the following sections 3.3 and 3.4, the state of impurities in the specimen is clarified and its influence on the interaction between a dislocation and the impurities will be described.

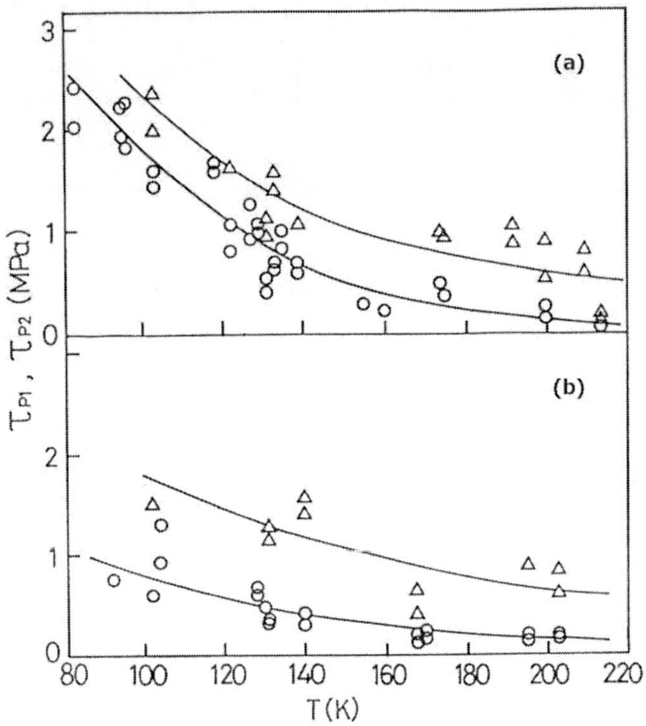

Figure 3-3. Dependence of (○) τ_{p1} and (△) τ_{p2} on temperature for (a) the quenched specimen and (b) the stored specimen.

3.3. Dependence of the State of Impurities for KCl:Sr^{2+} on Heat Treatment

Here is clarified the state of impurities in the specimen and is referred to the influence of the state of impurities on the interaction between a dislocation and the impurities, especially on the relation between temperature and the effective stress. The crystals used here are two kinds of specimens. The first is KCl:Sr^{2+} (0.050 mol.% in the melt) at the preceding section 3.1 and is hereafter named the quenched specimen. The second is prepared by keeping the quenched specimen at 370 K for 500 h and gradually cooling in a furnace for the purpose of aggregating impurities in it [4]. This is termed the annealed specimen.

Dielectric absorption of an I-V dipole causes a peak on the $\tan\delta$ -frequency relation. A relative formula which gives the proportionality between the concentration of I-V dipoles and a Debye peak height is expressed by [5]

$$\tan\delta = \frac{2\pi e^2 c}{3\varepsilon a k T}, \text{ (maximum)} \quad (3\text{-}1)$$

where e is the elementary electric charge, c is the concentration of the I-V dipole, ε is the dielectric constant in the matrix, a is the lattice constant, and kT has its usual meaning.

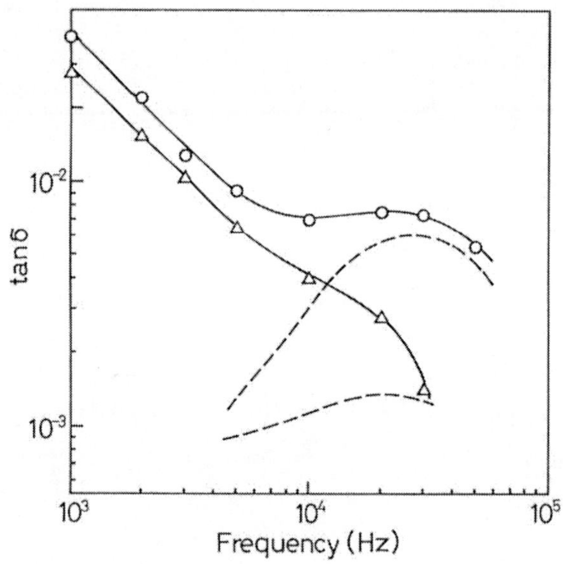

Figure 3-4. Dielectric loss in KCl containing Sr^{2+} ions at 393 K: (○) for the quenched specimen, (Δ) for the annealed specimen. (- - -) Losses coming from the dipoles.

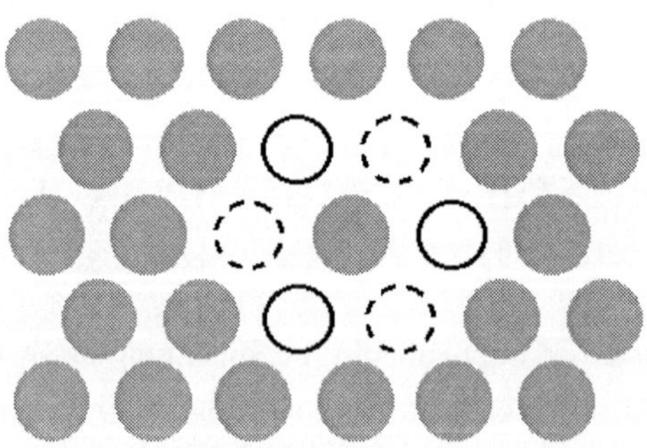

Figure 3-5. The trimer ((○) Sr^{2+}-(::) vacancy-(○) Sr^{2+}-(::) vacancy-(○) Sr^{2+}-(::) vacancy) structure proposed by Cook and Dryden [4]. A section through a (111) plane of the annealed KCl structure containing Sr^{2+}.

Figure 3-4 shows the influence of this heat treatment on the $\tan\delta$ -frequency curves for $KCl:Sr^{2+}$ at 393 K. The upper solid and dotted curves correspond to the quenched specimen and the lower curves the annealed specimen. Dotted lines show Debye peaks obtained by subtracting the d.c. part which is obtained by extrapolating the linear part of the solid curves in the low-frequency region to the high-frequency region. Introducing the peak heights of the dotted curves into equation (3-1), the concentration of the isolated dipole was determined to be 98.3 ppm for the quenched specimen and 21.8 ppm for the annealed specimen by dielectric loss measurement using an Andoh electricity TR-10C model. On the other hand, the atomic absorption gave 121.7 ppm for the Sr^{2+} concentration in the quenched specimen and 96.2 ppm for the annealed specimen. Therefore, it should be considered that 71.9 % of the I-V dipoles turn into the aggregates in KCl:Sr^{2+} single crystal and form at least trimers [4] by the heat treatment. The trimers have a structure in which three dipoles are arranged hexagonally head to tail in a (111) plane, as suggested by Cook and Dryden [4]. The close hexagonal structure of trimer (Sr^{2+}- vacancy- Sr^{2+}-vacancy-Sr^{2+}- vacancy) is shown in figure 3-5 [4].

3.4. Effective Stress and Critical Temperature

Figure 3-6 shows the variation of strain-rate sensitivity with stress decrement at a shear strain of 8 % for the annealed specimen at 125 K. The curve in the figure 3-6 has two bending points and two plateau regions at a given shear strain and temperature. The first plateau region ranges below the first bending point at low stress decrement and the second one extends from the second bending point at high stress decrement. The strain-rate sensitivity decreases with the stress decrement between the two bending points. The values of $\Delta\tau$ at the first and the second bending points are referred to as τ_{p1} and τ_{p2}, respectively. The same phenomena are observed for the quenched specimens as already shown in figures 2-10 and 2-11 of section 2.4. However, it is occasionally observed that the strain-rate sensitivity increases with stress decrement at the second plateau place on the curve for the annealed specimen. Figure 3-7 shows the phenomenon. The strain-rate sensitivity at second plateau place on the curve is considered to be due to dislocation cuttings (see sec. 2.7 of chap. 2). As for the quenched specimen, the same phenomenon as figure 3-7 does not appear.

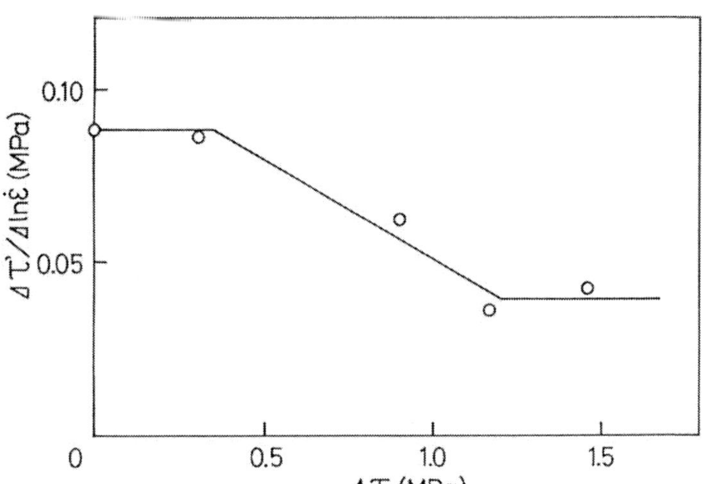

Figure 3-6. Relation between the strain-rate sensitivity and the stress decrement at a shear strain of 8 % for the annealed specimen at 125 K.

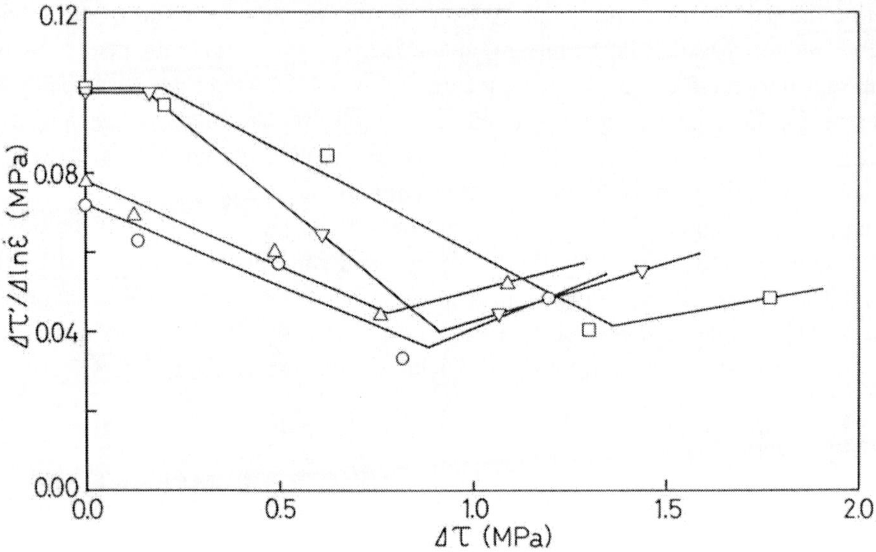

Figure 3-7. Relationship between the strain-rate sensitivity and the stress decrement for the annealed specimen at various conditions: (□) 183 K and ε =7 %, (▽) 187 K and ε =6 %, (○) 201 K and ε =6 %, (Δ) 201 K and ε =9 %.

The dependence of τ_{p1} and τ_{p2} on temperature for the quenched specimen is shown in figure 3-8 (a), while that for the annealed specimen is shown in figure 3-8 (b). The value of τ_{p1} becomes small by forming into the aggregates in the crystal and this result is clearer at lower temperature. This may be caused by the result that the separation between the weak obstacles lying on the mobile dislocation becomes wider as the I-V dipoles turn into aggregates. In addition, it is supposed that the decrease in τ_{p1} due to agglomerate of the I-V dipoles, i.e. softening, would be attributable to the loss of tetragonality in terms of the Fleischer's model [6], as suggested by Chin et al [7]. The decrease in the effective stress due to agglomerates of I-V dipoles has been reported for alkali halide crystals doped with divalent cations so far. Several experimental results are presented below.

1. Chin et al. have obtained the following experimental results in the paper (Chin, G. Y. et al. *J. Am. Ceram. Soc.* 1973, *56*, 369): KCl:Sr^{2+} (840 ppm) crystals soften on annealing at temperatures up to 773 K; the hardness of NaCl:Ca^{2+} (below 2000 ppm) crystals starts to increase rapidly after aging at 100 ℃ for 30 min and decreases on further annealing.
2. After quenching KCl:Sr^{2+} from 400 ℃, the measurements of the I-V dipoles and of the flow stress as a function of annealing time at 130 ℃ show that the concentration of isolated I-V dipoles decreases and the flow stress becomes lower at a longer annealing time (Dryden, J. S. et al. *Phil. Mag.* 1965, *12*, 379).
3. Although the agglomerates in the slowly cooled specimens (LiF:Mg^{2+} (75 ppm) crystals) act as stronger obstacles for mobile dislocations than isolated dipoles in the air-cooled specimens above 70 K, the specimens are softer in the slowly cooled (0.002 ℃ min^{-1}) than in the air-cooled (50 ℃ min^{-1}) condition below 70 K from the

temperature dependence of the critical resolved shear stress (Johnston, W. G. *J. Appl. Phys.* 1962, *33*, 2050).

4. The quenched specimens were prepared by quenching from 300 ℃ in an air stream to room temperature and the annealed specimens by annealing at 800 ℃ for 12 h with subsequent cooling for 12 h. Such a quenching procedure was employed in the paper (Johnston, W. G. *J. Appl. Phys.* 1962, *33*, 2050). As a result, LiF:Mg^{2+} (0.1 at.%) crystals are found to be softer below 120 K by the annealing (Gaiduchenya, V. F. et al. *Sov. Phys. Solid State* 1970, *12*, 27).
5. LiF:Mg^{2+} (0.4 mol.%) crystals soften also on annealing at 160 ℃ over 100 h (Buravleva, M. G. et al. *Sov. Phys. Solid State* 1980, *22*, 150).

As for τ_{p2}, no great difference is seen between the quenched specimen and the annealed one. Accordingly, as the I-V dipoles turn into the aggregates by the heat treatment, the difference between τ_{p1} and τ_{p2} obviously becomes wider at lower temperature. This suggests that the distribution of Sr^{2+} obstacles on mobile dislocation, i.e. the average length of dislocation segments for the annealed specimen is wider than that for the quenched one.

The critical temperature T_c at which τ_{p1} is zero is around 210 K for the annealed specimen. This T_c value is small in contrast to $T_c \approx 240$ K of the quenched specimen as can be seen from figures 3-8 (a) and (b).

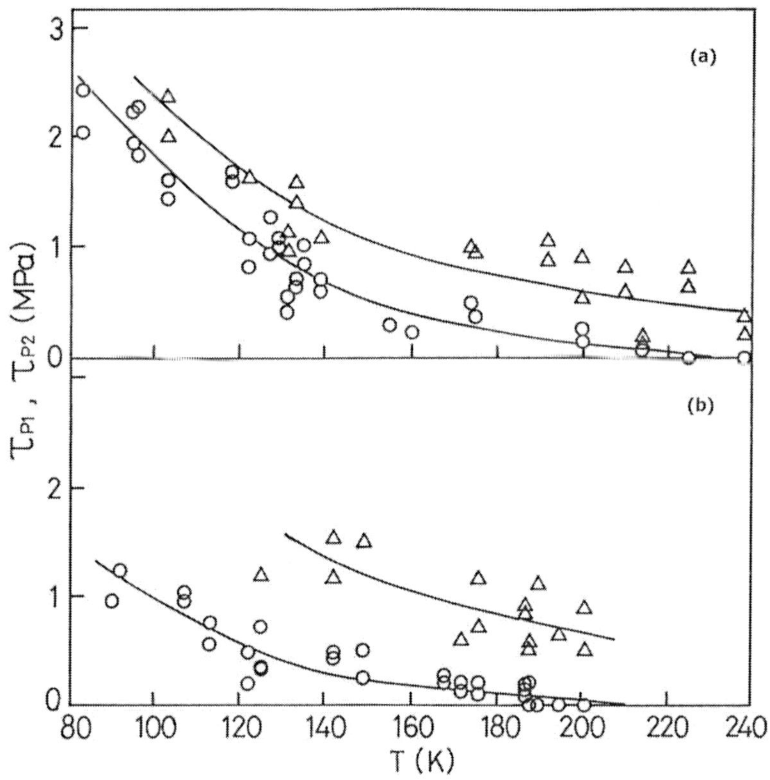

Figure 3-8. Dependence of (○) τ_{p1} and (Δ) τ_{p2} on temperature for KCl:Sr^{2+}: (a) for the quenched specimen and (b) for the annealed specimen.

CONCLUSION

The plots of the strain-rate sensitivity of flow stress and the stress decrement due to oscillation have two bending points and two plateau regions for the quenched specimen and similar result is observed also for the specimen stored at room temperature for a half year. On the basis of the relative curve of strain-rate sensitivity versus stress decrement, it was found that the effective stress, τ_{p1}, due to the impurities in the stored specimen is smaller than that of the quenched specimen within the temperatures.

The value of τ_{p1} becomes obviously smaller at low temperature when I-V dipoles turn into the aggregates (trimers) in $KCl{:}Sr^{2+}$. This may be caused by the results that the separation between the Sr^{2+} obstacles lying on the mobile dislocation becomes wider, in addition to the loss of tetragonality in terms of the Fleischer's model, as the I-V dipoles turn into the agglomerates. By forming into the agglomerates in $KCl{:}Sr^{2+}$ single crystals, the value of T_c also becomes small.

REFERENCES

[1] Green, M. L.; Zydzik, G. *Scripta Metall.* 1972, *6*, 991.
[2] Dryden, J. S.; Morimoto, S.; Cook, J. S. *Phil. Mag.* 1965, *12*, 379.
[3] Lubenets, S. V.; Startsev, V. I. *Soviet Phys. Solid State*, 1968, *10*, 15.
[4] Cook, J. S.; Dryden, J. S. *Proc. Phys. Soc.* 1962, *80*, 479.
[5] Lidiard, A. B. *Handbuch der Physik*, Springer; Berlin, 1957; Vol. 20, p 246.
[6] Fleischer, R. L. *J. Appl. Phys.* 1962, *33*, 3504.
[7] Chin, G. Y.; Van Uitert, L. G.; Green, M. L.; Zydzik, G. J.; Kometani, T. Y. *J. Amer. Ceram. Soc.* 1973, *56*, 369.

Chapter 4

EFFECTIVE STRESS AND CRITICAL TEMPERATURE OF KCL DOPED WITH VARIOUS DIVALENT CATIONS

4.1. INTRODUCTION

It has been reported in chapter 2 (see sec. 2.1) that the strain-rate cycling test during the Blaha effect measurement can separate the influence of impurities and forest dislocations on dislocation motion. The Blaha effect is the phenomenon whereby static flow stress decreases when an ultrasonic oscillatory stress is superimposed. On the basis of the measurement of strain-rate sensitivity under application of ultrasonic oscillatory stress during plastic deformation, the interaction between a dislocation and various divalent impurities is investigated in order to determine whether the divalent ionic size is an important factor for the solution hardening in this chapter 4. This is made by using four kinds of single crystals: KCl doped with Mg^{2+} (0.035 mol.% in the melt), Ca^{2+} (0.035 and 0.065 mol.% in the melt), Sr^{2+} (0.035, 0.050 and 0.065 mol.% in the melt) or Ba^{2+} (0.050 and 0.065 mol.% in the melt) as divalent impurities. The main aim of the present chapter is to estimate the effect of various divalent impurities on the critical temperature, at which the effective stress due to a weak obstacle lying on the dislocation is zero, and on the activation enthalpy for the break-away of a dislocation from the impurity. With respect to KCl:Sr^{2+} (0.050 mol.% in the melt), the influence of the state of divalent impurities on them is also described here for the annealed specimen used in chapter 3.

4.2. RELATION BETWEEN THE TEMPERATURE AND τ_{P1}, τ_{P2}, AND YIELD STRESS τ_Y

Figure 4-1 shows the influence of temperature on the relation between the strain-rate sensitivity, $\Delta\tau'/\Delta\ln\dot{\varepsilon}$, and the stress decrement, $\Delta\tau$, for KCl single crystals doped with Sr^{2+} as weak obstacles. As the temperature is high, the first bending point, τ_{p1}, shifts in the direction of low stress decrement and does not appear up to 225 K. The first plateau region indicates that the average length of the dislocation segment remains constant in that region. Thus, application of oscillation with low stress amplitude cannot influence the average length of the dislocation segment at low temperature, but even a low stress amplitude can do so at a temperature of 225 K. Such a phenomenon was also observed for the other specimens: KCl doped with Mg^{2+}, Ca^{2+} or Ba^{2+}.

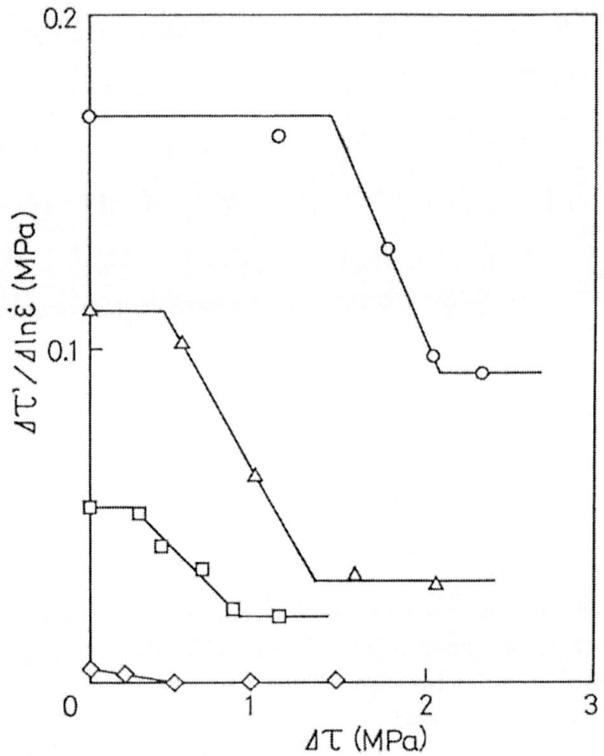

Figure 4-1. Relation between the strain-rate sensitivity and the stress decrement at the shear strain of 10 % for KCl:Sr^{2+} (0.050 mol.%) at various temperatures: (○) 103 K, (Δ) 133 K, (□) 200 K, (◇) 225 K.

Figures 4-2 (a) to (c) show the dependence of τ_{p1}, τ_{p2}, and yield stress (τ_y) on temperature for KCl:Sr^{2+} (0.035, 0.050 and 0.065 mol.%, respectively). It is clear from the figures that τ_{p1} and τ_{p2} tend to increase with decreasing temperature as well as τ_y for the three specimens and the curve of τ_y seems to approach a constant stress at high temperature. Both τ_{p1} and τ_{p2} increase with increasing Sr^{2+} concentration at a given temperature. Figure 4-3 shows the dependence of τ_{p1} and τ_{p2} on the yield stress for the specimen at 150 K. The plots correspond to the case when the impurity concentration is 0.035, 0.050, and 0.065 mol.% in the melt from the bottom. It can be seen from this figure that both τ_{p1} and τ_{p2} are approximately proportional to the yield stress. This means that τ_{p1} and τ_{p2} increase, depending on the Sr^{2+} concentration, because the yield stress generally becomes large with increasing impurity concentration [1-5]. Figures 4-4 to 4-6 show the relation between the three kinds of stress (i.e., τ_{p1}, τ_{p2}, and τ_y) and temperature for the other specimens. Similar results as the case of KCl:Sr^{2+} are also observed in these figures.

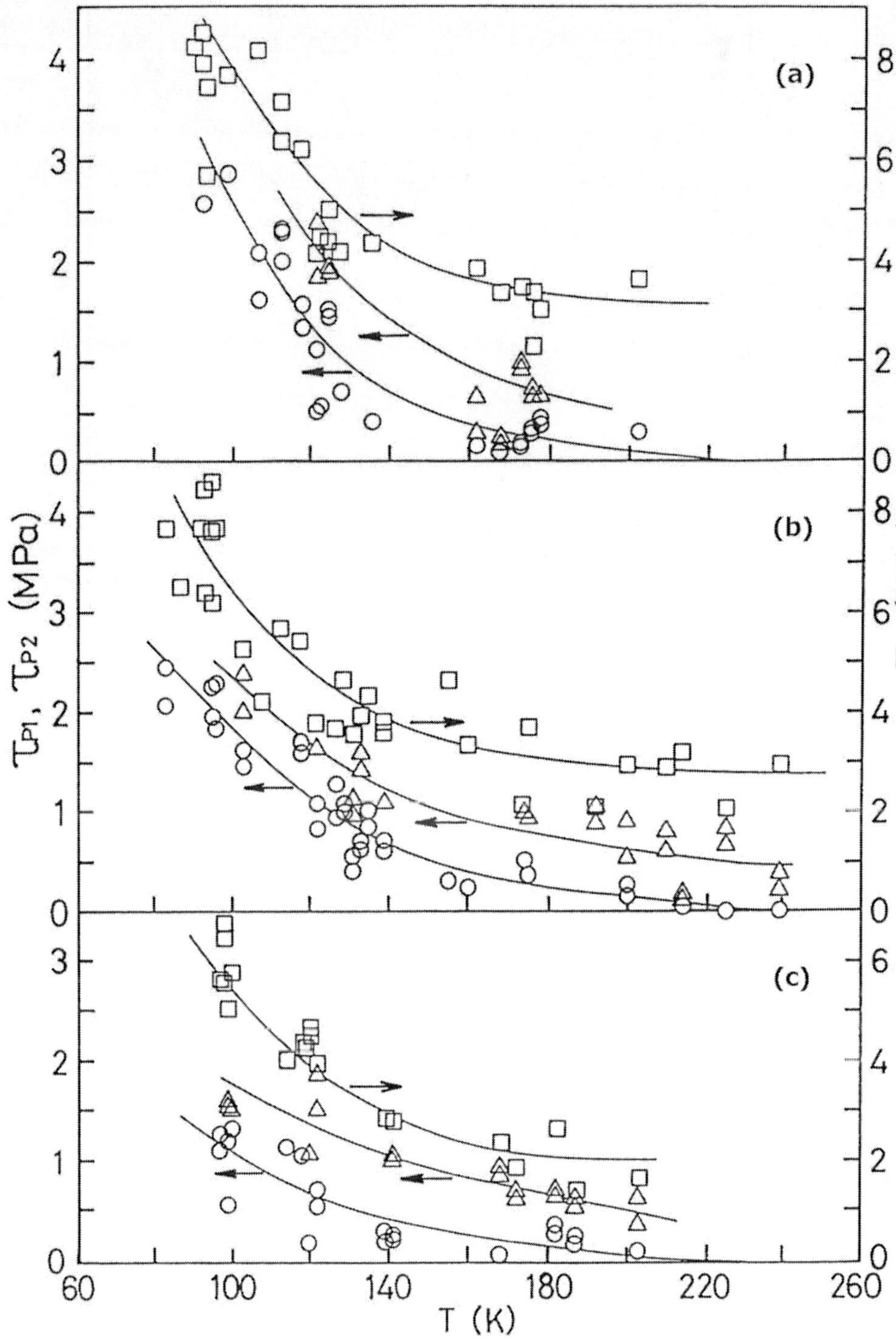

Figure 4-2. Dependence of (○) τ_{p1}, (Δ) τ_{p2}, and (□) τ_y on temperature for $KCl:Sr^{2+}$. Sr^{2+} concentration is (a) 0.065, (b) 0.050, and (c) 0.035 mol.%.

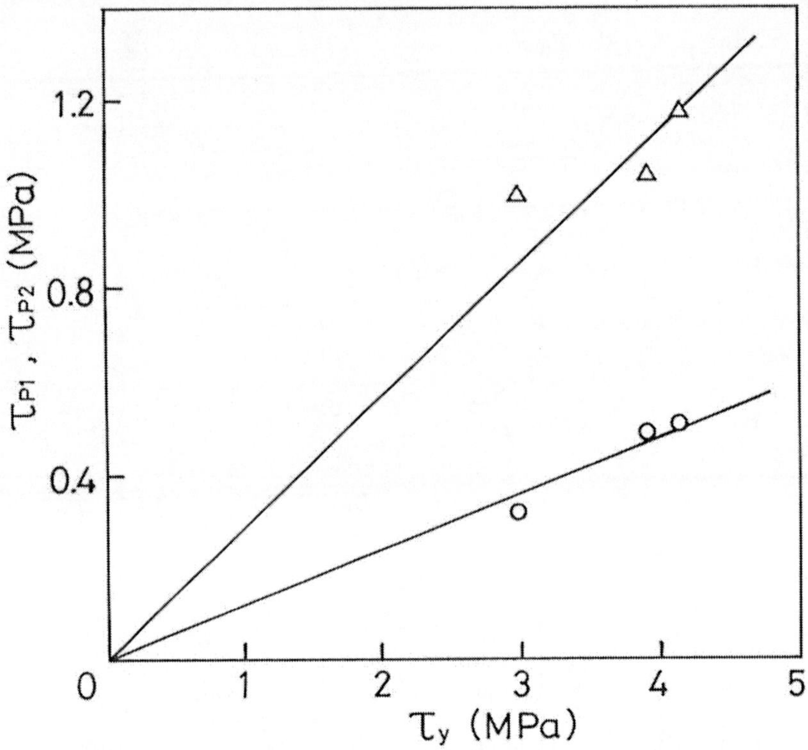

Figure 4-3. Dependence of (○) τ_{p1} and (Δ) τ_{p2} on the yield stress, τ_y, for KCl:Sr^{2+} at 150 K. Concentration of Sr^{2+} is 0.035, 0.050, 0.065 mol.% from the left.

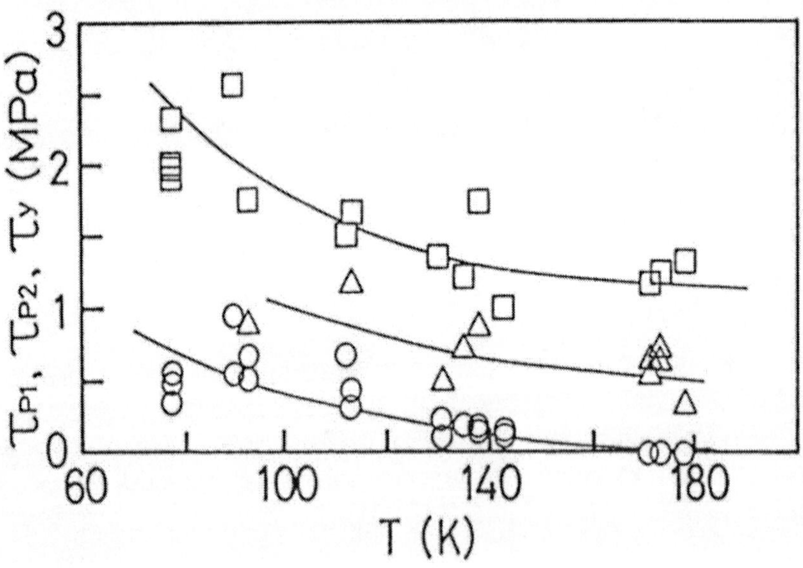

Figure 4-4. Dependence of (○) τ_{p1}, (Δ) τ_{p2}, and (□) τ_y on temperature for KCl:Mg^{2+} (0.035 mol.%).

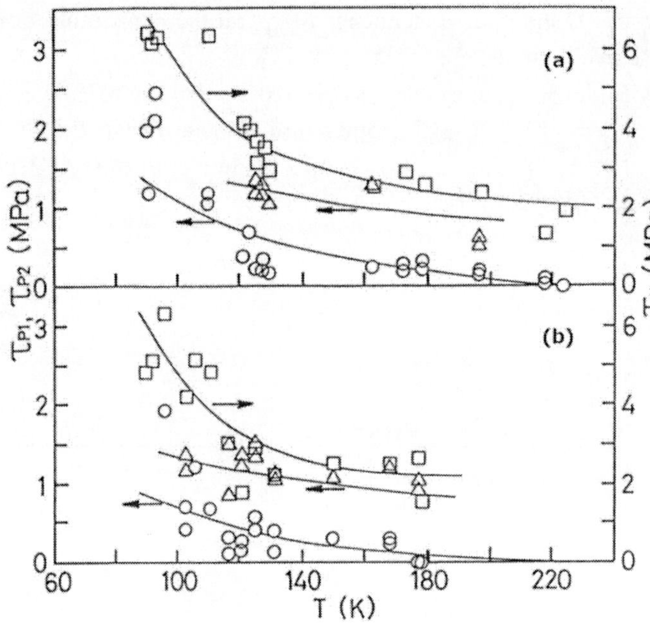

Figure 4-5. Dependence of (○) τ_{p1}, (Δ) τ_{p2}, and (□) τ_y on temperature for $KCl:Ca^{2+}$. Ca^{2+} concentration is (a) 0.065, (b) 0.050 mol.%.

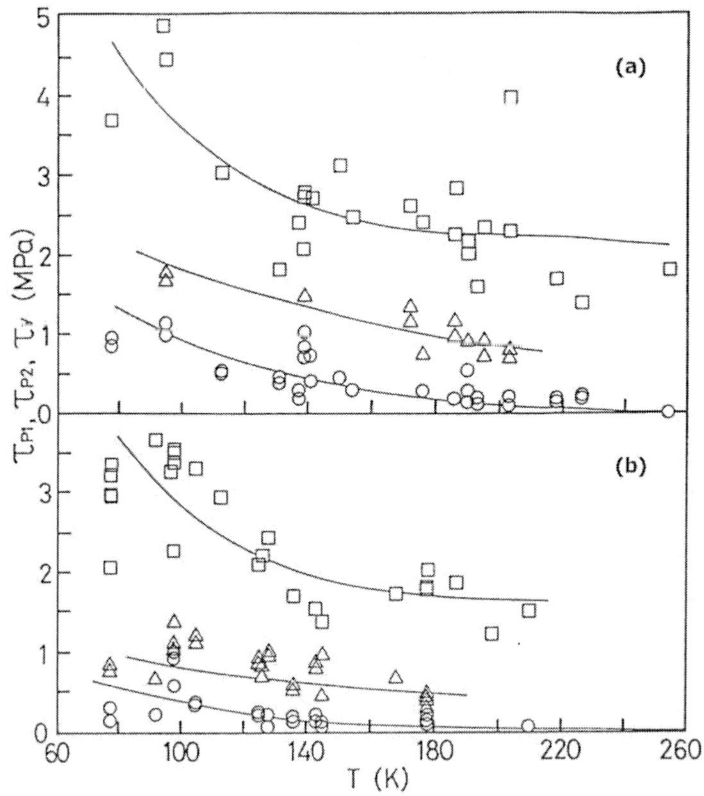

Figure 4-6. Dependence of (○) τ_{p1}, (Δ) τ_{p2}, and (□) τ_y on temperature for $KCl:Ba^{2+}$. Ba^{2+} concentration is (a) 0.065, (b) 0.050 mol.%.

Figures 4-7 (a) to (d) show the dependence of τ_{p1} on the temperature for KCl doped with Mg^{2+} (0.035 mol.% in the melt), Ca^{2+} (0.065 mol.% in the melt), Sr^{2+} (0.050 mol.% in the melt) or Ba^{2+} (0.065 mol.% in the melt). τ_{p1} decreases with increasing temperature for the four kinds of specimens. The critical temperature (T_c) at which the curves intersect the abscissa and τ_{p1} is zero, is around 180, 220, 230 and 260 K for KCl:Mg^{2+}, KCl:Ca^{2+}, KCl:Sr^{2+}, and KCl:Ba^{2+}, respectively.

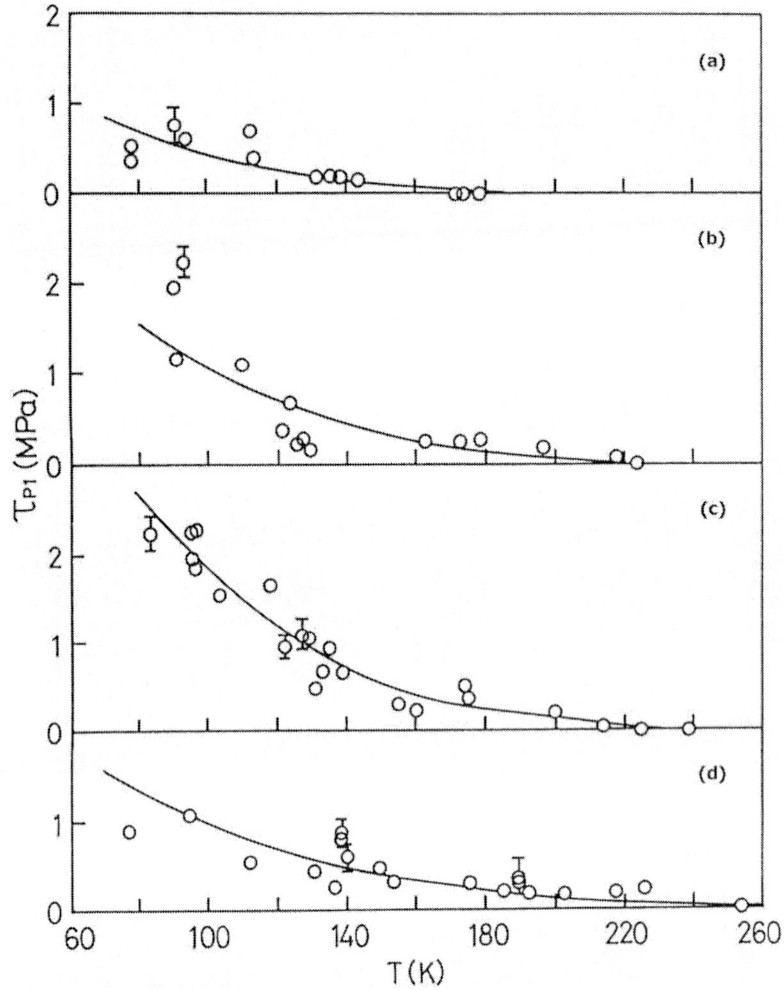

Figure 4-7. Dependence of τ_{p1} on the temperature for (a) KCl:Mg^{2+} (0.035 mol.%), (b) KCl:Ca^{2+} (0.065 mol.%), (c) KCl:Sr^{2+} (0.050 mol.%), and (d) KCl:Ba^{2+} (0.065 mol.%).

4.3. Critical Temperature

The tetragonal distortion resulting from the introduction of the divalent cations into alkali halides is generally formed in the lattice. A dislocation moves on a single slip plane and interacts strongly only with those defects lying within one atom spacing of the glide plane. It is well known that the Fleischer's model, which is the most successful one for the dramatic hardening of divalent cations, is suitable for the interaction between a dislocation and obstacles [1]. Then, the relation between the effective stress and the temperature can be approximated as

$$\left(\frac{\tau_{p1}}{\tau_{p0}}\right)^{1/2} = 1 - \left(\frac{T}{T_c}\right)^{1/2}, \tag{4-1}$$

where τ_{p0} is the effective stress due to the weak obstacles without thermal activation. The temperature dependence of τ_{p1} reveals the force-distance profile which expresses the interaction between a dislocation and weak obstacles. Figure 4-8 shows that the relation between τ_{p1} and temperature for KCl:Sr^{2+} agrees with the above equation.

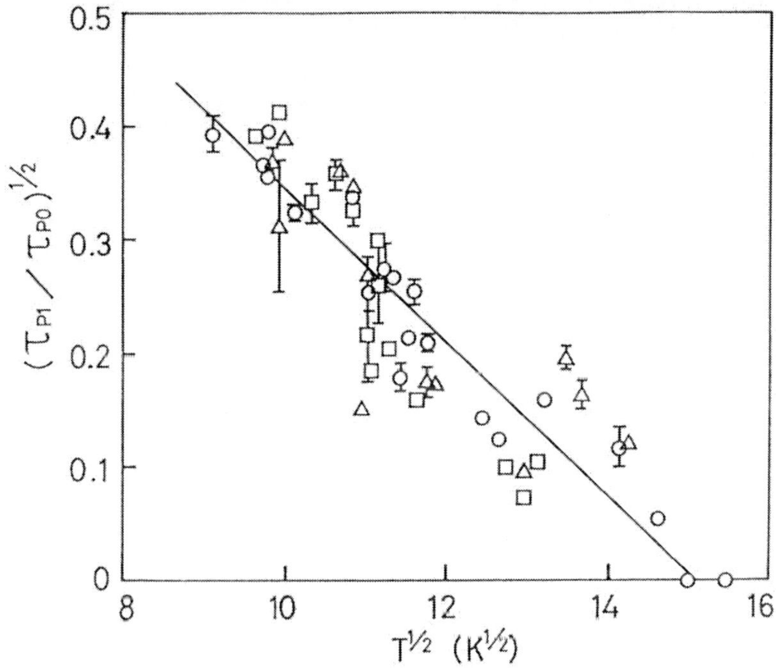

Figure 4-8. Linear plots of $(\tau_{p1}/\tau_{p0})^{1/2}$ and $T^{1/2}$ for KCl:Sr^{2+} (△) 0.035 mol.%, (○) 0.050 mol.%, and (□) 0.065 mol.%.

Table 4-1. Values of τ_{p0}.

KCl:Sr^{2+} (mol.%)	τ_{p0} (MPa)
0.035	9.01
0.050	14.52
0.065	16.71

The critical temperature can be also determined from figure 4-8. The value of the critical temperature, T_c, is about 227 K for KCl:Sr^{2+}, and τ_{p0}, which is obtained by extrapolating the relative line between $\tau_{p1}{}^{1/2}$ versus $T^{1/2}$ to 0 K, increases with Sr^{2+} concentration. The values of τ_{p0} are given in table 4-1.

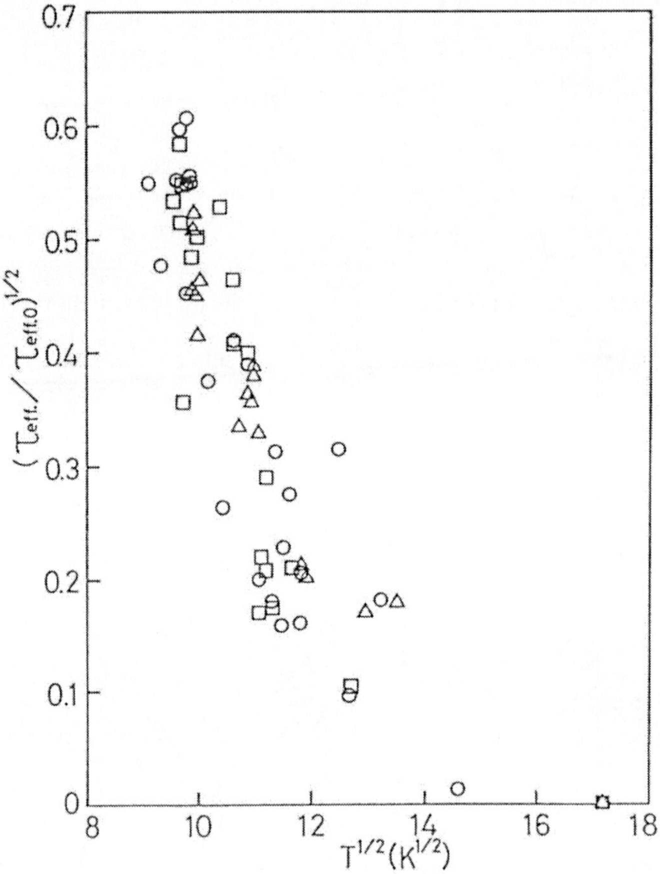

Figure 4-9. Relation between $(\tau_{eff.}/\tau_{eff.0})^{1/2}$ and $T^{1/2}$ for KCl:Sr^{2+} (△) 0.035 mol.%, (○) 0.050 mol.%, and (□) 0.065 mol.%. The yield stress at room temperature (about 295.5 K) is assumed to be the internal stress and $\tau_{eff.0}$ indicates the effective stress ($\tau_{eff.}$) exerted on the dislocation at 0 K.

Determination of T_c is attempted from the temperature dependence of yield stress. Then, the effective stress was defined by subtraction of the yield stress at room temperature from the yield stress at a given temperature. That is, the yield stress at room temperature is taken as the internal stress. The internal stress is 2.2, 3.2, and 3.7 MPa for KCl:Sr^{2+} (0.035 mol.%), Sr^{2+} (0.050 mol.%), and Sr^{2+} (0.065 mol.%), respectively. Figure 4-9 shows the relation between $(\tau_{eff.}/\tau_{eff.0})^{1/2}$ and $T^{1/2}$ for KCl:Sr^{2+}. $\tau_{eff.}$ shows the effective stress due to weak obstacle at a given temperature. $\tau_{eff.0}$ indicates the effective stress exerted on the dislocation at 0 K and is 16.6, 14.7, and 15.8 MPa for KCl:Sr^{2+} (0.035 mol.%), Sr^{2+} (0.050 mol.%), and Sr^{2+} (0.065 mol.%), respectively. As shown in the figure, the relation between $(\tau_{eff.}/\tau_{eff.0})^{1/2}$ and $T^{1/2}$ for KCl:Sr^{2+} cannot be approximated to a straight line. The value of internal stress in this assumption may be doubtful because the yield stress will be gradually decreased over room temperature from figures 4-2 and 4-4 to 4-6.

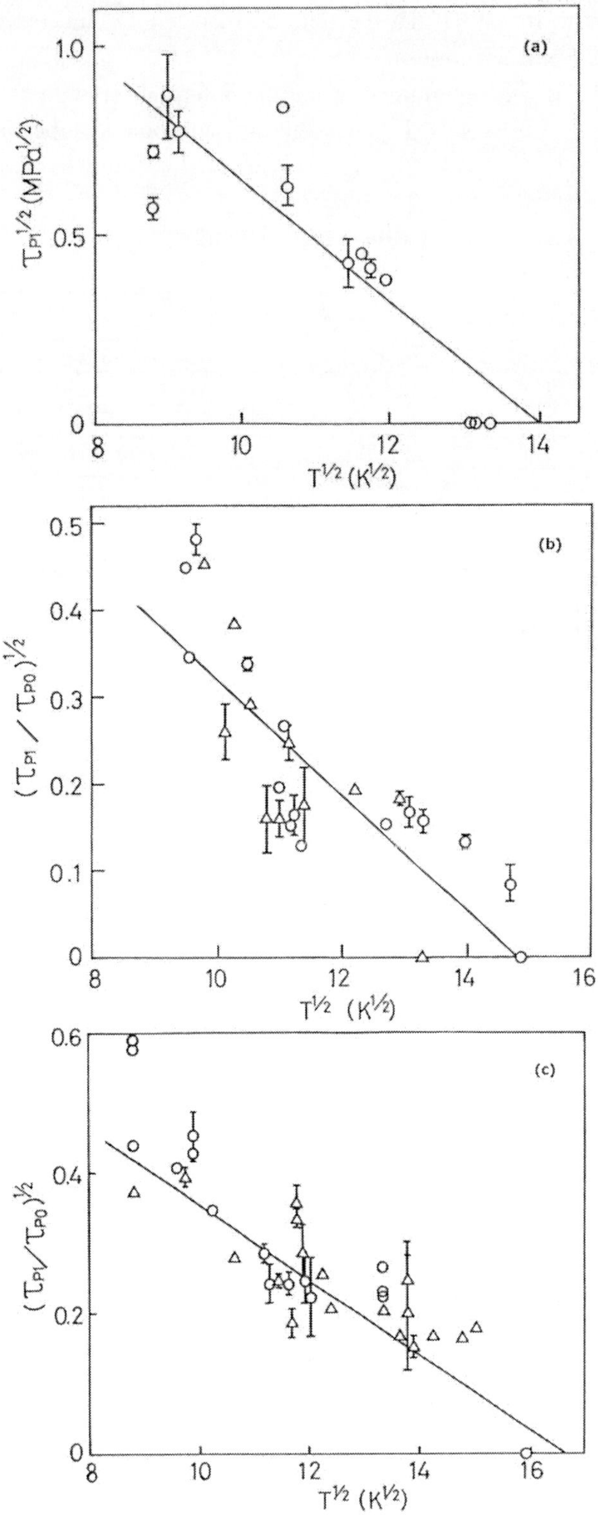

Figure 4-10. Linear plots of $\tau_{p1}^{1/2}$ and $T^{1/2}$ for (a) KCl:Mg^{2+} (○) 0.035 mol.% and of $(\tau_{p1}/\tau_{p0})^{1/2}$ and $T^{1/2}$ for (b) KCl:Ca^{2+} (Δ) 0.035 mol.% and (○) 0.065 mol.%, (c) KCl:Ba^{2+} (○) 0.050 mol.% and (Δ) 0.065 mol.%.

Figures 4-10 (a) to (c) show that the relation between τ_{p1} and temperature agrees with equation (4-1) for several specimens. The values of τ_{p0} are given in table 4-2. Then, the critical temperature can be determined from the intersection with the abscissa for each specimen and are shown in table 4-3. When the divalent ionic size becomes closer to that of K^+ from the small divalent cationic size side as illustrated in figure 4-11, T_c tends to increase.

Table 4-2. Values of τ_{p0}

Single crystal	τ_{p0} (MPa)
KCl:Mg^{2+} (0.035 mol.%)	5.64
KCl:Ca^{2+} (0.035 mol.%)	8.14
(0.065 mol.%)	9.63
KCl:Ba^{2+} (0.050 mol.%)	3.81
(0.065 mol.%)	6.82

Table 4-3. Values of T_c and ionic radius

Single crystal	T_c (K)	ionic radius (Å)	
KCl:Mg^{2+}	191	Mg^{2+}	0.72
KCl:Ca^{2+}	221	Ca^{2+}	1.00
KCl:Sr^{2+}	227	Sr^{2+}	1.13
KCl:Ba^{2+}	277	Ba^{2+}	1.36
		K^+	1.38

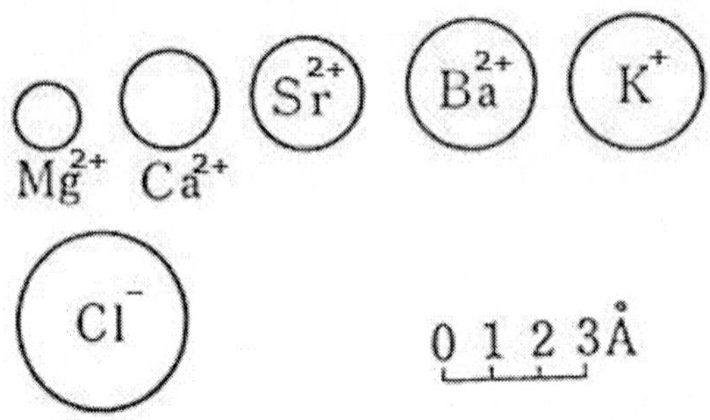

Figure 4-11. Illustration of ionic size.

As for the annealed specimen (KCl:Sr^{2+} (0.050 mol.%)), the relation between τ_{p1} and temperature is as follows. Two models will be considered here. One is the Fleischer's model and the other is the model [2] in which the force-distance profile is a triangle. The triangle profile gives the following relation:

$$\left(\frac{\tau_{p1}}{\tau_{p0}}\right)^{2/3} = 1 - \left(\frac{T}{T_c}\right)^{1/2}. \qquad (4\text{-}2)$$

Figure 4-12 shows whether equations (4-1) and (4-2) satisfy the data. If the interaction between a dislocation and aggregates for KCl:Sr^{2+} can be approximated to the Fleischer's model, τ_{p0} is 8.9 MPa and T_c is 223 K. However, if the force-distance profile is a triangle, τ_{p0} is 5.3 MPa and T_c is 211 K. As can be seen from the figure, it is difficult to determine whether the relation between the effective stress and temperature for the annealed specimen can be approximated to the Fleischer's model within the temperature range.

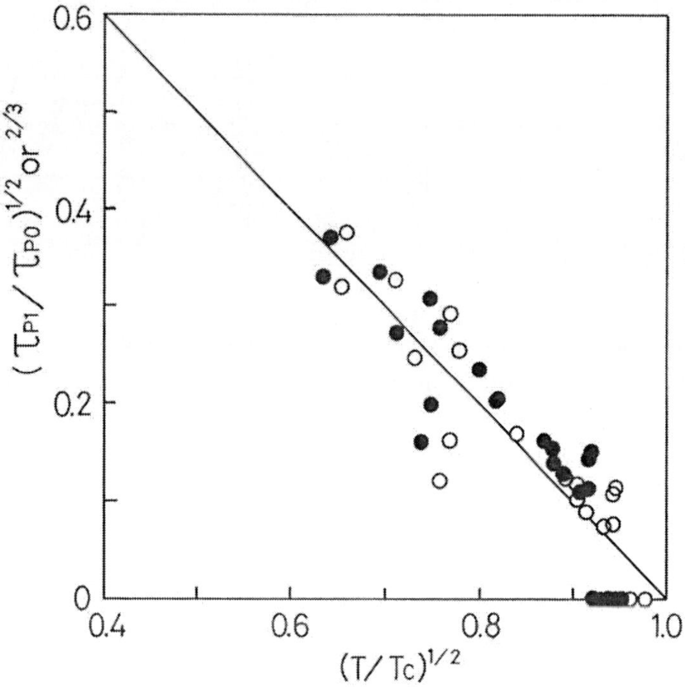

Figure 4-12. The relation between effective stress and temperature for the annealed specimen. (●) $(\tau_{p1}/\tau_{p0})^{1/2}$ - $(T/T_c)^{1/2}$; (○) $(\tau_{p1}/\tau_{p0})^{2/3}$ - $(T/T_c)^{1/2}$.

4.4. Relation between the Strain-Rate Sensitivity and the Stress Decrement

The strain-rate cycling test associated with the ultrasonic oscillation can separate the effective stress due to weak obstacles such as dopant ions from that due to dislocation cuttings. A general relation between the strain-rate sensitivity and the stress decrement due to oscillation at a given strain is schematically drawn in figure 4-13, where the curve has two bending points and two plateau regions. The phenomenon shown in the figure reflects the influence of ultrasonic oscillation on the dislocation motion on the slip plane containing many weak obstacles (dopant ions) and a few strong ones (forest dislocations) during plastic deformation. The means seems useful until the forest dislocation density is below about 10^{10} cm^{-2}. The reason is as follows.

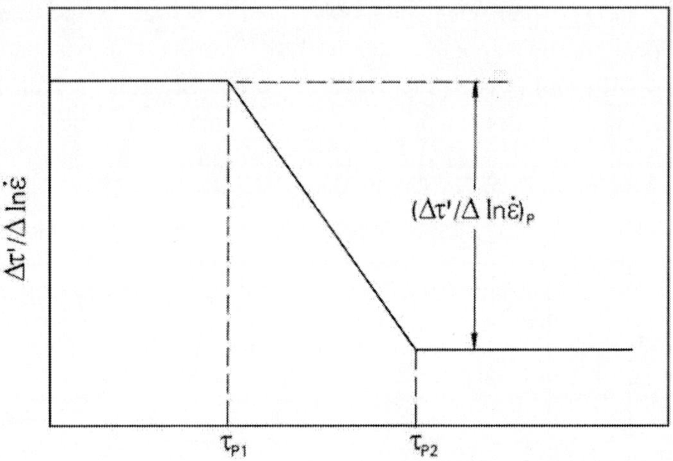

Figure 4-13. Illustration of relationship between the strain-rate sensitivity of flow stress and the stress decrement at a given strain.

Assuming that the average spacing of impurities along the dislocation, l_P, is nearly equal to that of impurities on the slip plane, the concentration of the impurities, c, is expressed by

$$c \propto \frac{1}{l_P^2}. \tag{4-3}$$

When the impurity ionic radius is 2 Å and c is 100 ppm, is found to be 400 Å through equation (4-3). The average length of dislocation segments pinned by forest dislocations, l_f, is 3200 Å on the assumption that dislocation density (ρ) is 10^9 cm^{-2}. In the case that ρ is 10^{10} cm^{-2}, decreases to 1000 Å. As for ρ at 10^9 cm^{-2}, therefore, it is imaged that dislocations are pinned by many impurities between forest dislocations as given in figure 4-14 (a). Figure 4-14 (b) shows the aspect of a dislocation for ρ at 10^{10} cm^{-2}. Only a few impurities impede the dislocation motion between forest dislocations. Accordingly, strain-rate cycling tests during the Blaha effect measurement seem to be recommendable to the study on the interaction between a dislocation and the impurities in the specimens in which ρ is below 10^{10} cm^{-2}.

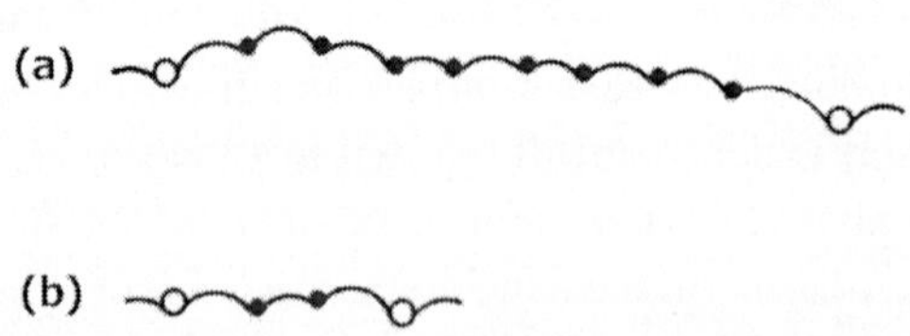

Figure 4-14. Aspect of a dislocation pinned by impurities between forest dislocations for the case: (a) ρ is 10^9 cm^{-2} and (b) ρ is 10^{10} cm^{-2}. A solid circle represents impurity and an open circle forest dislocations.

The first bending point, τ_{p1}, has been regarded as the effective stress due to the weak obstacles which lie on the mobile dislocation, when the dislocation moves forward with the aid of thermal activation. Therefore, observation of τ_{p1} provides information on the interaction between a dislocation and weak obstacles. At the stress decrement of more than τ_{p2}, the obstacles to the dislocation motion are only strong ones such as forest dislocations and the dopant ions no longer have an act as obstacles. The strain-rate sensitivity due to weak obstacles, $(\Delta\tau'/\Delta\ln\dot{\varepsilon})_p$, is defined in figure 4-13. It is assumed from the following two more experimental results in addition to the description in the section 2.1 (introduction) of chapter 2 that $(\Delta\tau'/\Delta\ln\dot{\varepsilon})_p$ is the strain-rate sensitivity due to impurities.

1. Figures 4-15 and 4-16 show the relative curves of the strain-rate sensitivity and the stress decrement at several strains for KCl:Ba^{2+} (0.050 mol.%) at 178 K and KCl:Sr^{2+} (0.050 mol.%) at 200 K. The first phenomenon is that the curve of figures 4-15 and 4-16 shift upwards with increasing shear strain because the strain-rate sensitivity due to dislocation cuttings increases with increasing strain, while $(\Delta\tau'/\Delta\ln\dot{\varepsilon})_p$ is constant independently of the strain.
2. The other phenomenon is that the $(\Delta\tau'/\Delta\ln\dot{\varepsilon})_p$ is proportional to the square root of impurity concentration (i.e. $c^{1/2}$), which is related to the inverse of the average length, l, of the dislocation segments (viz. $(\Delta\tau'/\Delta\ln\dot{\varepsilon})_p = kT/bld$, where $l = l_p$, b is the magnitude of Burgers vector, d is the activation distance, and kT has its usual meaning), at a given temperature. This is shown in figure 4-17. The concentration of the I-V (Impurity-Vacancy) dipole is estimated by the dielectric loss measurement.

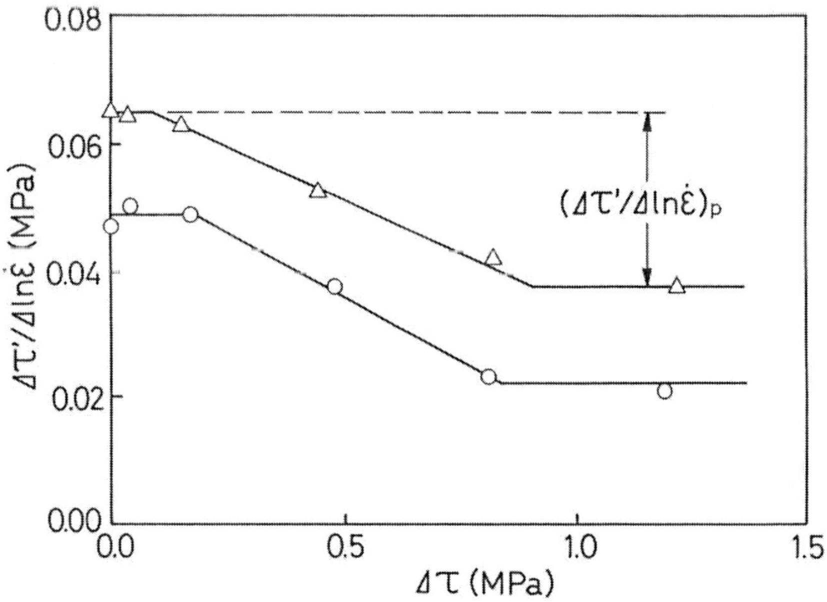

Figure 4-15. Strain-rate sensitivity versus stress decrement curve for KCl:Ba^{2+} (0.050 mol.%) at 178 K. ε : (○) 10 %, (Δ) 13 %. $(\Delta\tau'/\Delta\ln\dot{\varepsilon})_p$ is assumed to be the strain-rate sensitivity due to impurities.

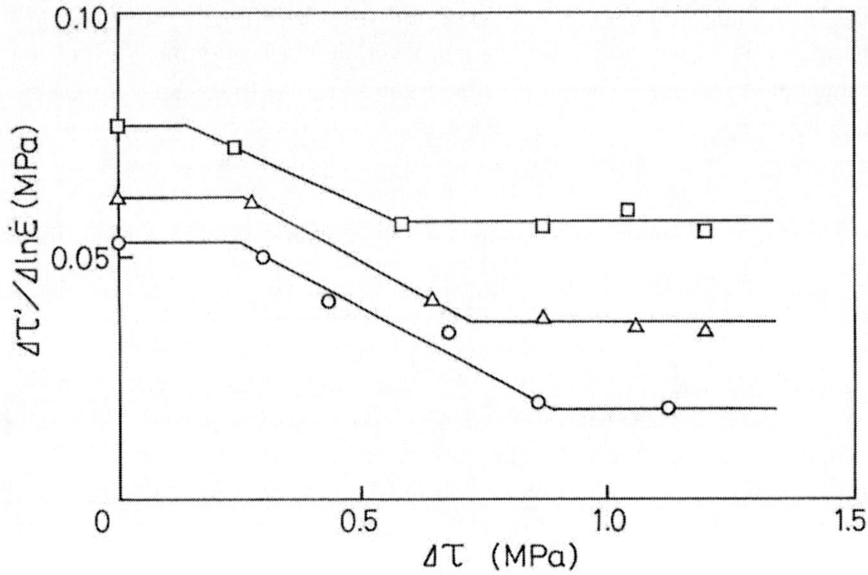

Figure 4-16. Strain-rate sensitivity versus stress decrement curve for KCl:Sr^{2+} (0.050 mol.%) at 200 K. ε : (○) 10 %, (Δ) 14 %, and (□) 18 %.

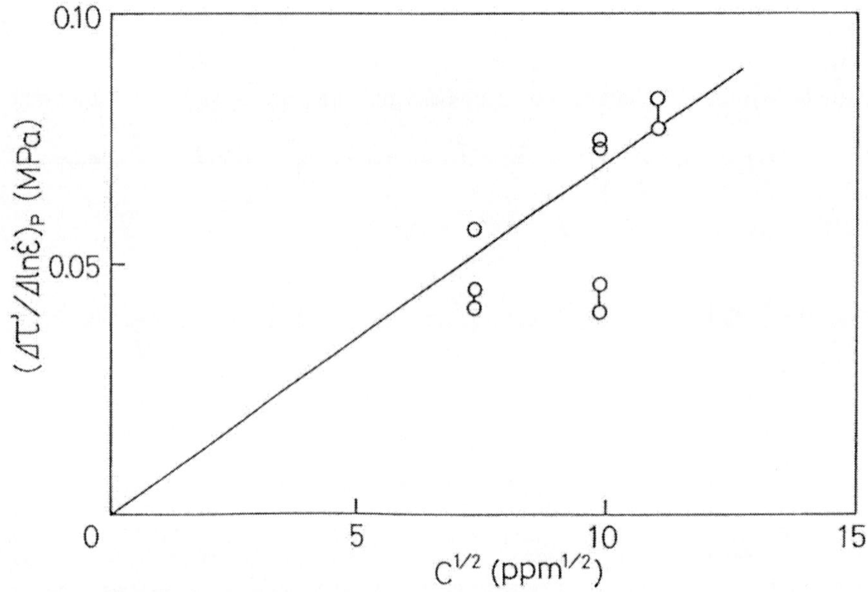

Figure 4-17. Relationship between $(\Delta\tau'/\Delta\ln\dot{\varepsilon})_p$ and $c^{1/2}$ for KCl:Sr^{2+} at 120 to 130 K.

4.5. Activation Energy for the Interaction between a Dislocation and the Divalent Impurity

The rate-controlling mechanism during plastic deformation of KCl doped with divalent impurities is thermally activated overcoming of the weak obstacles dispersed in the lattice by dislocation motion [3]. Then the strain rate ($\dot{\varepsilon}$) at low temperature is assumed to be given by an Arrhenius-type relationship [4]:

$$\dot{\varepsilon} = \dot{\varepsilon}_0 \exp\left(\frac{-\Delta G}{kT}\right), \tag{4-4}$$

where $\dot{\varepsilon}_0$ is a frequency factor which is unique for a particular dislocation mechanism, ΔG is the change in Gibbs free energy of activation for the dislocation motion, and kT has the usual meaning. If changes in entropy (ΔS) are neglected, ΔG of equation (4-4) may be replaced by the activation enthalpy (ΔH). Assuming that a dislocation moves at a fixed velocity, the activation enthalpy is expressed by

$$\Delta H = \alpha kT, \tag{4-5}$$

where α is a constant. When a dislocation can overcome weak obstacles dispersed in the lattice with the aid of thermal fluctuations, the values of the activation enthalpy are given by [5-8]

$$\Delta H = -kT^2 \left(\frac{\partial \ln \dot{\varepsilon}}{\partial \tau^*}\right)_T \left(\frac{\partial \tau^*}{\partial T}\right)_{\dot{\varepsilon}}, \tag{4-6}$$

where $(\partial \ln \dot{\varepsilon} / \partial \tau^*)_T$ is obtained from $(\Delta \tau' / \Delta \ln \dot{\varepsilon})_p$ at a given temperature. $(\partial \tau^* / \partial T)_{\dot{\varepsilon}}$ in equation (4-6) can be obtained from the differentiation of equation (4-1) as follows

$$\left(\frac{\partial \tau_{p1}}{\partial T}\right)_{\dot{\varepsilon}} = \left\{1 - \left(\frac{T}{T_c}\right)^{-1/2}\right\} \frac{\tau_{p0}}{T_c}. \tag{4-7}$$

The values of the activation enthalpy derived using equation (4-6) are plotted as a function of temperature in figures 4-18 (a) to (c). It is seen from these figures that the activation enthalpy is proportional to the temperature in accordance with the requirements of equation (4-5). The value of $\Delta H(T_c)$, taken from figures 4-18 (a) to (c), corresponds to the activation enthalpy for the overcoming of the defects by a dislocation at 0 K. The values are 0.50, 0.63, and 0.71 eV for KCl:Ca^{2+}, KCl:Sr^{2+}, and KCl:Ba^{2+}, respectively. The values of the activation enthalpy for KCl:Mg^{2+} could not be derived because the second bending point, τ_{p2}, in figure 4-13 did not appear in the experiment. The results that $\Delta H(T_c)$ increases with increasing divalent cationic size suggest that the tetragonality, $\Delta\varepsilon$, resulting from the formation of the I-V dipole for alkali halides doped with divalent cations in the Fleischer's model ($F_0 = \mu \Delta \varepsilon b^2 / 3.86$, where F_0 is the maximum force on the dislocation towards the defect, μ is the shear modulus, and b is the slip vector of the dislocation) [1, 9] increases when the divalent ionic size increasingly approaches that of the K^+ ion, as illustrated in figure 4-11.

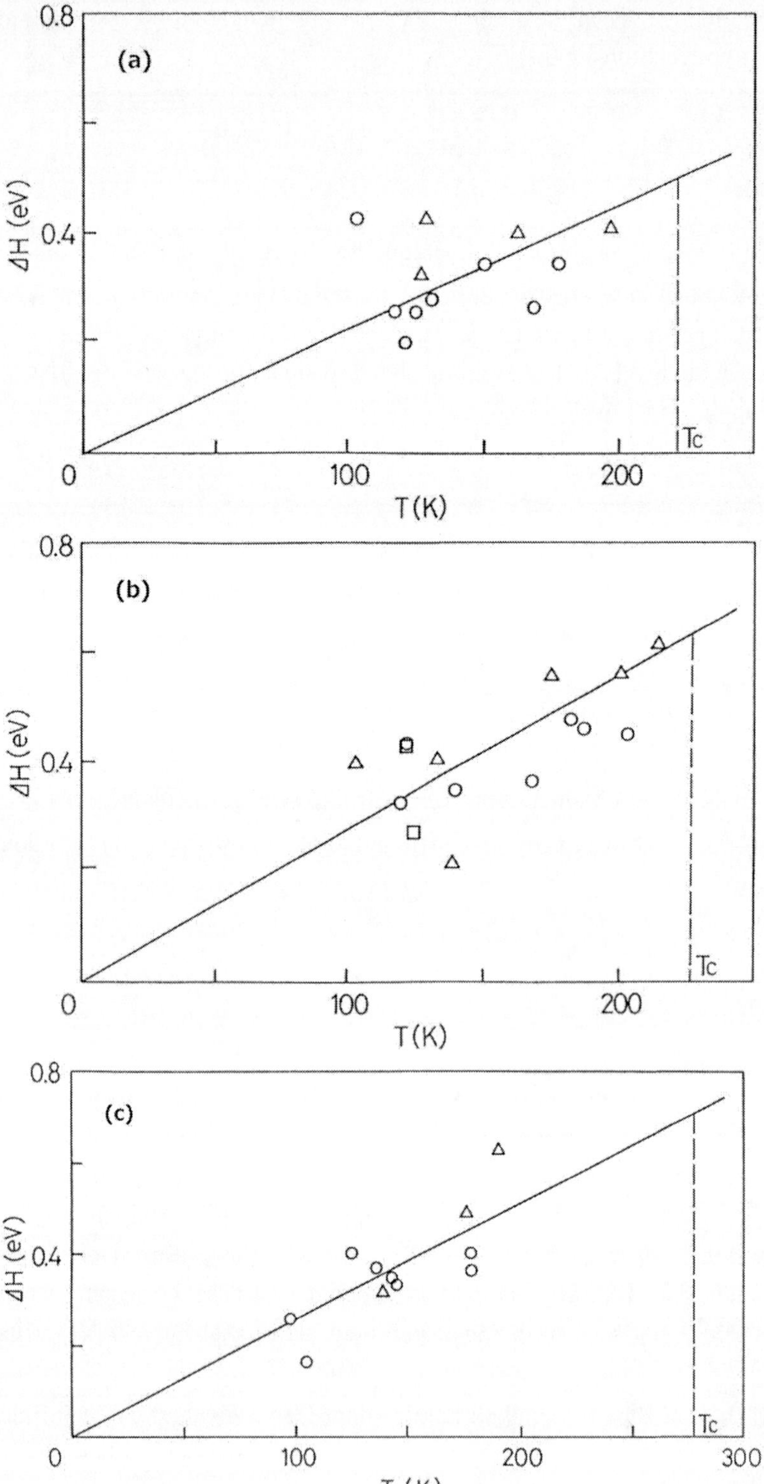

Figure 4-18. Proportional relationship between the temperature and the activation enthalpy for (a) KCl:Ca^{2+} (○) 0.035 mol.% and (△) 0.065 mol.%, (b) KCl:Sr^{2+} (○) 0.035 mol.%, (△) 0.050 mol.%, and (□) 0.065 mol.%, and (c) KCl:Ba^{2+} (○) 0.050 mol.% and (△) 0.065 mol.%.

The relation between the temperature and activation enthalpy for the annealed specimen ($KCl:Sr^{2+}$ (0.050 mol.%)) is shown in figure 4-19. Assuming the interaction between a dislocation and weak obstacles (the agglomerate I-V dipoles) can be approximated by the Fleischer's model, the temperature and activation enthalpy relation corresponds to the solid circles in figure 4-19. However, the data do not appear to be satisfied by equation (4-5). On the other hand, plots of open circles in figure 4-19 show the activation enthalpy obtained using equation (4-2). The activation enthalpy is then nearly proportional to temperature for plots of open circles, compared to plots of solid circles as shown in figure 4-19. In this case, the value of $\Delta H(T_c)$ for the annealed specimen ($KCl:Sr^{2+}$) taken from figure 4-19 is 0.37 eV. Thus the force-distance profile between a dislocation and aggregate for $KCl:Sr^{2+}$ can be approximated by a triangle rather than the Fleischer's model. Consequently, the interaction between a dislocation and the agglomerate of I-V dipoles in $KCl:Sr^{2+}$ cannot be approximated by the Fleischer's model.

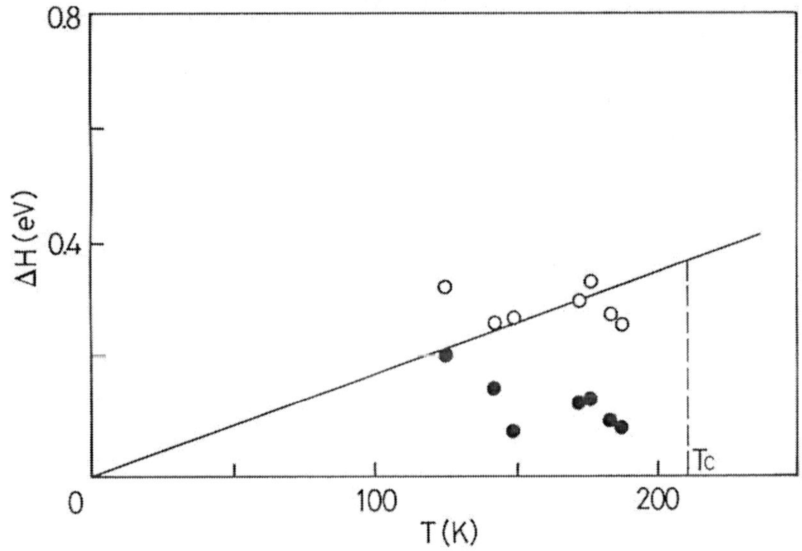

Figure 4-19. Proportional relationship between the temperature and the activation enthalpy for the annealed specimen. (●) $(\tau_{p1}/\tau_{p0})^{1/2}$ - $(T/T_c)^{1/2}$; (○) $(\tau_{p1}/\tau_{p0})^{2/3}$ - $(T/T_c)^{1/2}$.

Conclusion

The T_c for each specimen was derived from the relation between $T^{1/2}$ and $\tau_{p1}^{1/2}$ which reveals the force-distance curve between a dislocation and weak obstacle with respect to the Fleischer's model. T_c tends to increase when the divalent cationic size is increasingly close to the K^+ ionic size in KCl single crystals doped with Mg^{2+}, Ca^{2+}, Sr^{2+} or Ba^{2+} as divalent impurities. Furthermore, the values of the activation enthalpy for the interaction between a dislocation and the impurity are 0.50, 0.63, and 0.71 eV for $KCl:Ca^{2+}$, $KCl:Sr^{2+}$, and $KCl:Ba^{2+}$, respectively. The results suggest that the tetragonality around the dipole increases with $\Delta H(T_c)$ and T_c from the Fleischer's model.

Analyzing the data obtained from the strain-rate cycling tests associated with the ultrasonic oscillation for two kinds of single crystals: the quenched $KCl:Sr^{2+}$ specimens and the annealed ones, the following three results were mainly found on the basis of linear plots of effective stress versus temperature and proportionality of temperature versus activation enthalpy. First, the force-distance profile between a dislocation and the weak obstacle cannot be approximated by the Fleischer's model when I-V dipoles turn into aggregates (at least trimers) by the heat treatment. Secondly, the values of τ_{p0} and the critical temperature (T_c), at which the effective stress is zero, for the annealed specimen is smaller in contrast to that of the quenched specimen. Thirdly, comparing with the quenched specimen, the activation enthalpy $\Delta H(T_c)$ for the break-away of a dislocation from the impurity also becomes small for the annealed specimen.

REFERENCES

[1] Fleischer, R. L. *J. Appl. Phys.* 1962, *33*, 3504.
[2] Foreman, A. J. E.; Makin, M. J. *Phil. Mag.* 1966, *14*, 911.
[3] Guiu, F.; Langdon, T. G. *Phil. Mag.* 1974, *30*, 145.
[4] Christian, J. W.; Masters, B. C. *Proc. R. Soc. A* 1964, *281*, 240.
[5] Conrad, H. *J. Metals* 1964, *16*, 582.
[6] Conrad, H. *J. Phys.* 1967, *45*, 581.
[7] Dey, B. N.; Tyson, W. R.; *Phys. Status Solidi* (a) 1972, *9*, 215.
[8] Kataoka, T.; Uematsu, T.; Yamada, T. *Jpn. J. Appl. Phys.* 1978, *17*, 271.
[9] Fleischer, R. L. *Acta Metall.* 1962, *10*, 835.

Chapter 5

THE FRIEDEL RELATION AND THERMALLY ACTIVATED FLOW PARAMETERS OF KCL DOPED WITH VARIOUS DIVALENT CATIONS

5.1. INTRODUCTION

When alkali halide crystals are doped with divalent cations, the ions are expected to be paired with positive ion vacancies. The pairs are termed I-V dipoles and strongly interact with screw as well as edge dislocations. It has been mentioned for KCl single crystals doped with Mg^{2+}, Ca^{2+}, Sr^{2+} or Ba^{2+} that the interaction between a dislocation and the various divalent impurities can be approximated to the Fleischer's model [1] in the previous chapters. So far, however, the interaction between a dislocation and the impurities is examined for the specimen without discussing whether the Friedel relation can be taken into the Fleischer's model.

It is well known that the Friedel relation [2] between the effective stress and the average length of dislocation segments can be applied to most weak obstacles to dislocation motion at low solute concentration. Using a digital computer, the variation of the critical shear stress with the bending angle at which the dislocation breaks-away from the obstacle has been determined for the range of 0 to 180 degrees by Foreman and Makin [3]. As a result, it was found within the accuracy of computations that the application of the Friedel relation is allowed when the included angle between the arms of the dislocation at the obstacle is above approximately 140 degrees.

As for the interaction between a dislocation and the impurity in KCl: Mg^{2+}, Ca^{2+}, Sr^{2+} or Ba^{2+} single crystals, whether the Friedel relation can be taken into the Fleischer's model or not is a matter for argument. This is examined in this chapter 5. Furthermore, activation energy for break-away of the dislocation from the impurity is obtained for the each specimen.

5.2. RELATION BETWEEN $(L_0/L)^2$ AND Φ_0

Now we imagine that a mobile dislocation encounters a linear array of weak obstacles (A, B, and C) on the slip plane. Although the mobile dislocation is impeded by them, it breaks away from the obstacle B with the help of thermal activation and meets another one, B'. It reaches AB'C. The process is illustrated in figure 5-1. When the dislocation ABC (solid line) moves to AB'C (dashed one), the sweeping area (A_0) is given by

$$A_0 = L_0^2 \approx hL\,, \tag{5-1}$$

where h is the distance of advance and L is the average spacing of impurities along a dislocation. L_0 is that of impurities on a slip plane and is obtained by [1, 4]

$$L_0 = \frac{b}{\left(\frac{4c}{3}\right)^{1/2}}\,, \tag{5-2}$$

where b is the magnitude of the Burgers vector and c is the concentration of the obstacles. The shaded area in figure 5-1 corresponds to A_0. The force F acting on the dislocation ABC is given by

$$F = \tau^* bL \tag{5-3}$$

and also

$$F = 2E\sin\theta\,, \tag{5-4}$$

where τ^* is the effective shear stress, E is the line tension of the dislocations, and θ is the angle at which the dislocation breaks away from the obstacle. When the interaction between the dislocation and the obstacles is weak, namely θ is small, the A_0 is expressed by [5]

$$A_0 = L_0^2 = L^2\theta\,. \tag{5-5}$$

Furthermore, we can find

$$\sin\theta = (L/2)/R = \frac{h}{L} \tag{5-6}$$

and then

$$\theta = \frac{h}{L}\,, \tag{5-7}$$

where R is the radius of curvature. Equation (5-5) is named the Fleischer relation [5] and can be rewritten as follows:

$$\left(\frac{L_0}{L}\right)^2 = \frac{\pi - \phi}{2}, \tag{5-8}$$

where ϕ is the angle at which the dislocation embraces the obstacle [6, 7]. At the temperature of 0 K, equation (5-8) is expressed as

$$\left(\frac{L_0}{L}\right)^2 = \frac{\pi - \phi_0}{2}. \tag{5-9}$$

Substituting of equation (5-6) in equation (5-4) gives

$$F = \frac{2Eh}{L}. \tag{5-10}$$

Combining equations (5-3) and (5-10), we find

$$L^2 = \frac{2Eh}{\tau^* b}. \tag{5-11}$$

Further combining equations (5-1) and (5-11) gives

$$L = \left(\frac{2L_0^2 E}{\tau^* b}\right)^{1/3}. \tag{5-12}$$

Equation (5-12) is termed the Friedel relation [2].

The ϕ_0 under the effective shear stress is given by (see the subsequent sec. 5.4 of this chap. 5)

$$\cos\left(\frac{\phi_0}{2}\right) = \left(\frac{\tau^* L_0 b}{2E}\right)^{2/3}. \tag{5-13}$$

From equations (5-12) and (5-13), it follows that

$$\left(\frac{L_0}{L}\right)^2 = \cos\left(\frac{\phi_0}{2}\right). \tag{5-14}$$

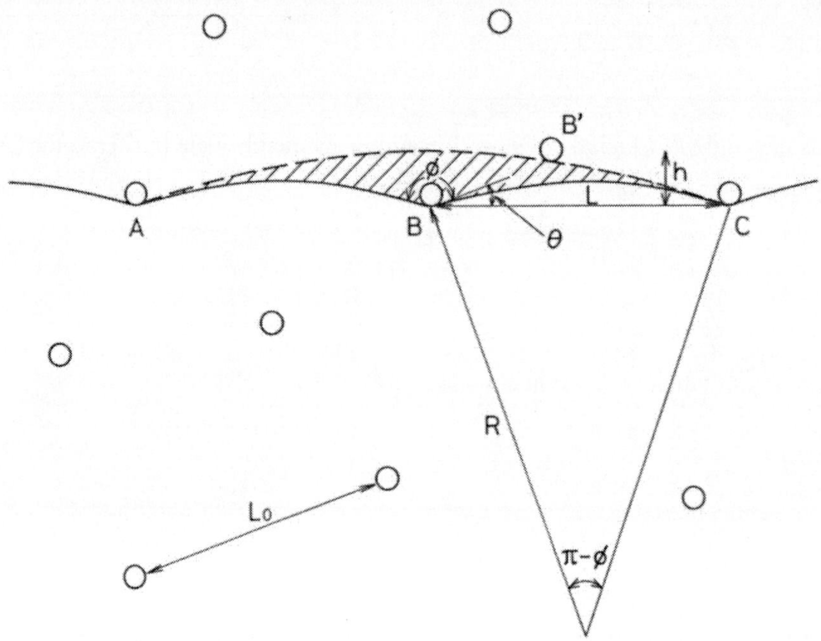

Figure 5-1. Dislocation held up at weak obstacles on the slip plane. Open circles represent one type of weak obstacles, such as impurities. The shaded area corresponds to the sweeping area of the mobile dislocation when the dislocation ABC moves forward to AB'C by the free flight motion.

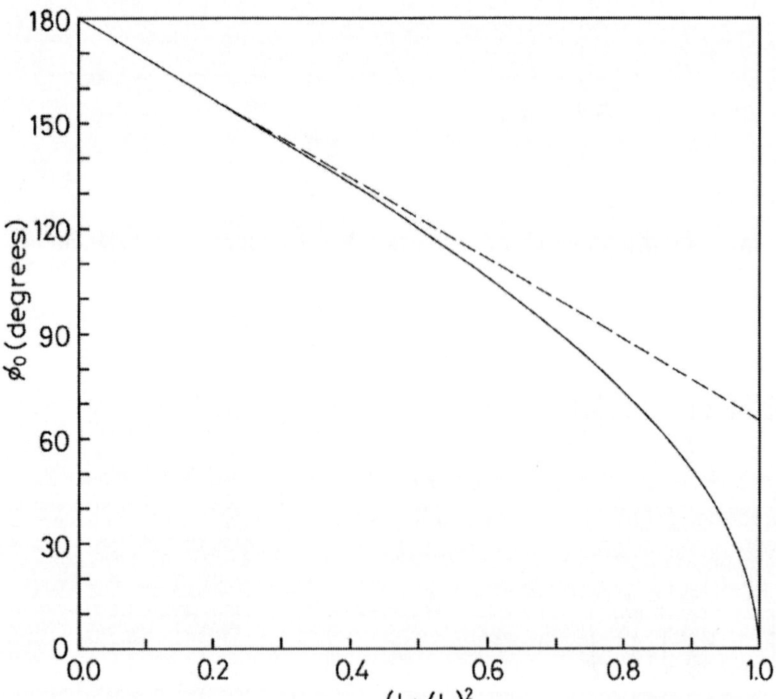

Figure 5-2. Relationship between $(L_0 / L)^2$ and ϕ_0. Solid line represents the curve obtained from the Friedel relation (see equation (5-14)); dashed line from the Fleischer relation (see equation (5-9)).

In figure 5-2, the relation between $(L_0/L)^2$ and ϕ_0 for equation (5-9) and that for equation (5-14) are represented as the dashed and solid lines, respectively. As mentioned above, equation (5-9) is derived from the Fleischer relation (i.e. equation (5-5)) which was suggested by Fleischer separately from the Friedel relation [2]. Equation (5-14) is based on the Friedel relation (i.e. equation (5-12)) proposed originally by Friedel. The solid line shows agreement with the dashed one above ϕ_0 of about 150 degrees and L/L_0 is above about 1.9 from figure 5-2. On the other hand, the ϕ_0 for the solid line is less than that for the dashed one at a given $(L_0/L)^2$ below ϕ_0 of about 150 degrees. The difference between values of ϕ_0 for the two kinds of lines obviously becomes wider with increasing the $(L_0/L)^2$.

The Fleischer relation implies that the angle through which the dislocation is bent by the obstacle A or C becomes the same θ as the case of obstacle B while dislocation ABC moves to AB'C in figure 5-1. Considering that dislocation AB'C successively breaks away also from obstacles A and C, the Fleischer relation is in essential agreement with the Friedel relation that the steady state is reached, i.e. a dislocation steadily overcomes thermal obstacles during steady and plastic deformation [6, 7]. It can be realized that the weak obstacles suitable for the Friedel relation will be simultaneously satisfied with the Fleischer relation. In other words, the applicable range of ϕ_0 in equation (5-14) agrees with that of ϕ_0 in equation (5-9). This leads to the result that the solid line overlaps with the dashed one in figure 5-2. Therefore, it is considered that the Friedel relation, as well as the Fleischer one, is applicable to weak obstacles such as impede the dislocation within $150° \leq \phi_0 \leq 180°$.

5.3. Relation between Effective Shear Stress and Temperature

Fleischer [1] approximated the force-distance relation between a dislocation and the divalent impurity for LiF:Mg^{2+} (0.008 mol.%) by

$$F = \frac{F_0}{\left(\frac{x}{b}+1\right)^2} \tag{5-15}$$

and found the Gibbs free energy (ΔG) for the interaction between a dislocation and the divalent impurity by the following equation:

$$\Delta G = \int_0^B \left\{ F_0 \Big/ \left(\frac{x}{b}+1\right)^2 - F \right\} dx, \tag{5-16}$$

where F_0 is the F at the temperature of 0 K, x is the distance from the defect and B is the value of x at which F equals the defect force. Then, F and F_0 are expressed by using equation (5-12) as follows

$$F = \left(2L_0^2 E\right)^{1/3} b^{2/3} \tau^{*2/3}, \qquad (5\text{-}17)$$

$$F_0 = \left(2L_0^2 E\right)^{1/3} b^{2/3} \tau_0^{*2/3}, \qquad (5\text{-}18)$$

where τ^*_0 is the effective shear stress at 0 K. From equation (5-12) and (5-15), B in equation (5-16) is given by

$$B = b\left\{\left(\frac{\tau_0^*}{\tau^*}\right)^{1/3} - 1\right\}. \qquad (5\text{-}19)$$

Substituting of equations (5-17) to (5-19) in equation (5-16) gives

$$\Delta G = \Delta G_0 \left\{1 - \left(\frac{\tau^*}{\tau_0^*}\right)^{1/3}\right\}^2, \ (\Delta G_0 = F_0 b) \qquad (5\text{-}20)$$

and also from an Arrhenius equation for the thermally activated deformation rate ($\dot{\varepsilon}$) the Gibbs free energy of activation is expressed as

$$\Delta G = \alpha k T, \ \left(\alpha = \ln\left(\frac{\dot{\varepsilon}_0}{\dot{\varepsilon}}\right)\right) \qquad (5\text{-}21)$$

where kT has the usual meaning and $\dot{\varepsilon}_0$ is a frequency factor. From the substitutional equation of equation (5-21) in equation (5-20), the relation of effective shear stress and temperature for the Fleischer's model taking account of the Friedel relation is expressed in the form:

$$\left(\frac{\tau^*}{\tau_0^*}\right)^{1/3} = 1 - \left(\frac{T}{T_c}\right)^{1/2}, \ \left(T_c = \frac{\Delta G_0}{\alpha k}\right). \qquad (5\text{-}22)$$

This model is hereafter termed the F-F. Therefore, the relative formula of the effective shear stress, τ_{p1}, due to the impurities and temperature for the F-F is

$$\left(\frac{\tau_{p1}}{\tau_{p0}}\right)^{1/3} = 1 - \left(\frac{T}{T_c}\right)^{1/2}. \tag{5-23}$$

τ_{p1} depends on temperature and on the type and density of impurities, and is considered to represent the effective stress due to only one type of the impurities which lie on the dislocation when the dislocation moves forward with the help of oscillation as described in the chapters 2 and 3. The results which agree with equation (5-23) are shown in figures 5-3 (a) to (d) for KCl: Mg^{2+}, Ca^{2+}, Sr^{2+} or Ba^{2+} single crystals. The value of τ_{p0}, which is obtained by extrapolating linear relationship between $\tau_{p1}^{1/3}$ and $T^{1/2}$ to 0 K, is given in table 5-1 and increases with the concentration of impurities for KCl:Sr^{2+}. No difference between τ_{p0} for KCl:Ca^{2+} (0.035 mol.% in the melt) and for KCl:Ca^{2+} (0.065 mol.% in the melt) can be observed in the table. This is because the concentration (c) of divalent impurities in KCl:Ca^{2+} (0.035 mol.% in the melt), of which c is determined to be 43.1 ppm, is nearly equal to that for KCl:Ca^{2+} (0.065 mol.% in the melt), of which c is 43.5 ppm (see table 5-3).

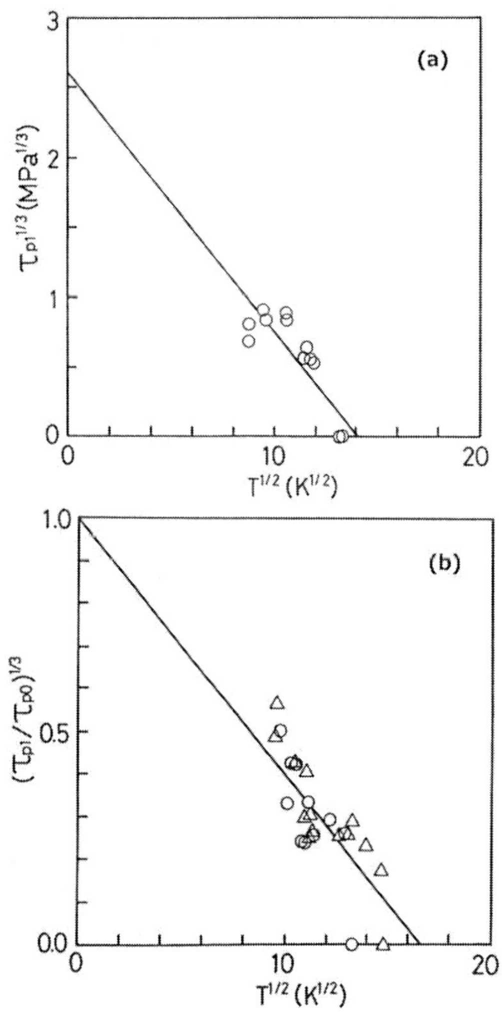

Figure 5-3. (Continued).

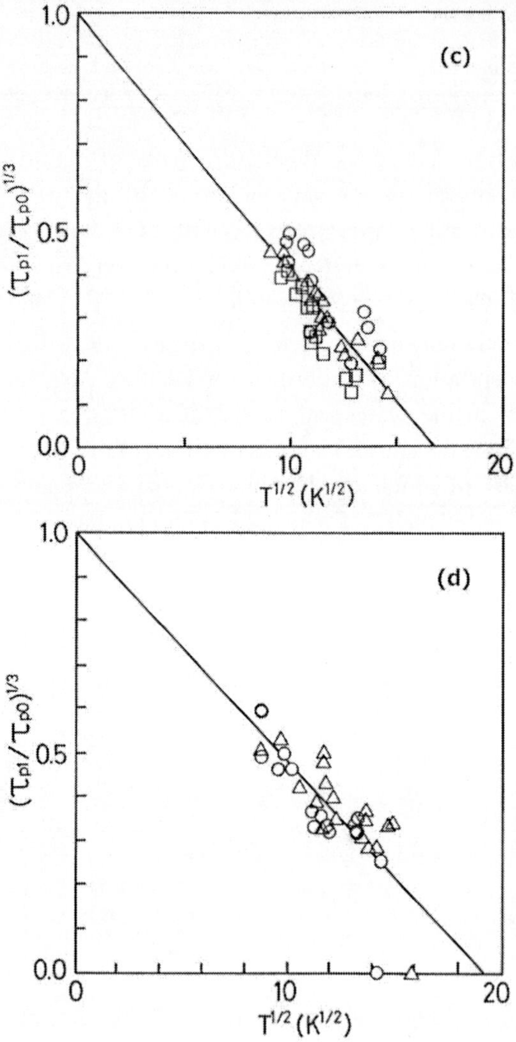

Figure 5-3. Relationship between the effective stress and the temperature at the F-F for (a) KCl:Mg^{2+} (○) 0.035 mol.% (b) KCl:Ca^{2+} (○) 0.035 mol.% and (Δ) 0.065 mol.% and (c) KCl:Sr^{2+} (○) 0.035 mol.%, (Δ) 0.050 mol.%, and (□) 0.065 mol.% and (d) KCl:Ba^{2+} (○) 0.050 mol.% and (Δ) 0.065 mol.%.

Table 5-1. Values of τ_{p0} for the F-F

Single crystal (mol.%)	τ_{p0} (MPa)
KCl:Mg^{2+} (0.035)	17.91
KCl:Ca^{2+} (0.035)	15.34
(0.065)	14.11
KCl:Sr^{2+} (0.035)	11.44
(0.050)	25.47
(0.065)	36.31
KCl:Ba^{2+} (0.050)	5.34
(0.065)	7.27

From the value at which the relative line of $\tau_{p1}^{1/3}$ versus $T^{1/2}$ intersects the abscissa, T_c can be determined as shown in table 5-2. T_c tends to increase with increasing the divalent cationic size.

Table 5-2. Values of T_c for the F-F

Single crystal	T_c (K)
$KCl:Mg^{2+}$	199
$KCl:Ca^{2+}$	278
$KCl:Sr^{2+}$	289
$KCl:Ba^{2+}$	370

Table 5-3. Values of c and ϕ_0

Single crystal (mol.%)	c (ppm)[a]	ϕ_0 (degrees)
$KCl:Mg^{2+}$ (0.035)	7.9	130
$KCl:Ca^{2+}$ (0.035)	43.1	155
(0.065)	43.5	157
$KCl:Sr^{2+}$ (0.035)	55.2	161
(0.050)	98.3	154
(0.065)	121.8	149
$KCl:Ba^{2+}$ (0.050)	9.2	159
(0.065)	28.3	163

a The concentration of the impurities is obtained from dielectric loss measurement.

5.4. Bending Angle of a Dislocation Pinned by the Impurity

Under the effective shear stress (τ_{p0}) at 0 K, the bending angle (ϕ_0) at which a dislocation embraces the impurity is obtained as below

$$2E\cos\left(\frac{\phi_0}{2}\right) = \tau_{p0}Lb \tag{5-24}$$

and

$$L = \left(\frac{2L_0^2E}{\tau_{p0}b}\right)^{1/3} \tag{5-25}$$

namely,

$$\frac{\phi_0}{2} = \cos^{-1}\left(\frac{\tau_{p0} L_0 b}{2E}\right)^{2/3}, \tag{5-26}$$

where the line tension of the dislocations is calculated by μb^2. The shear modulus (μ) for the [110] direction is assumed to be 1.01×10^{10} Pa at 0 K [8]. The average spacing of impurities on the slip plane is given by equation (5-2), where the concentration of the impurities is obtained from dielectric loss measurement for each specimen. The calculations of equation (5-26) are tabulated in table 5-3. Therefore, since the values of ϕ_0 for KCl: Ca^{2+}, Sr^{2+} or Ba^{2+} single crystals exceed about 150 degrees, it was confirmed that the F-F mentioned in section 5.3 is appropriate for the interaction between a dislocation and the impurity in the specimens. However, it is seemed not to be appropriate for KCl:Mg^{2+}, because the ϕ_0 is below 150 degrees. Consequently, the Friedel relation is supposed to be applied to KCl: Ca^{2+}, Sr^{2+} or Ba^{2+}, but not to KCl:Mg^{2+} although the Friedel relation is exact for most weak obstacles to dislocation motion at low concentration of the obstacles.

If the force-distance relation between a dislocation and the impurity in KCl:Mg^{2+} is approximated to the Fleischer's model, the values of T_c and τ_{p0} are 191 K and 5.64 MPa, which can be found by linear plots of $\tau_{p1}^{1/2}$ and $T^{1/2}$.

5.5. Dependence of the Strain-Rate Sensitivity Due to the Impurities on the Temperature

When a dislocation overcomes the impurity with the aid of thermal fluctuations, the activation enthalpy (ΔH) is given by the relation [9-12]:

$$\Delta H = -kT^2\left(\frac{\partial \ln \dot{\varepsilon}}{\partial \tau^*}\right)_T\left(\frac{\partial \tau^*}{\partial T}\right)_{\dot{\varepsilon}}. \tag{5-27}$$

The $(\partial \ln \dot{\varepsilon}/\partial \tau^*)_T$ in equation (5-27) is obtained from a reciprocal of the $(\Delta\tau'/\Delta\ln\dot{\varepsilon})_p$ which is given by the difference between strain-rate sensitivity at first plateau region and at second one on the relative curve of strain-rate sensitivity and stress decrement. The reason why $(\Delta\tau'/\Delta\ln\dot{\varepsilon})_p$ is the strain-rate sensitivity due to impurities was already mentioned in section 2.1 of chapter 2 and section 4.4 of chapter 4. The ΔH can also be expressed as the following equation on the assumption that the changes in entropy are neglected [9, 11]

$$\Delta H = \alpha k T. \tag{5-28}$$

Combining equations (5-27) and (5-28), we find

$$\left(\frac{\partial \tau^*}{\partial \ln \dot{\varepsilon}}\right)_T = -\left(\frac{\partial \tau^*}{\partial T}\right)_{\dot{\varepsilon}} \frac{T}{\alpha}. \tag{5-29}$$

In equation (5-29), the $(\partial \tau^* / \partial T)_{\dot{\varepsilon}}$ for the F-F can be expressed from the differentiation of equation (5-23) with respect to the temperature by

$$\frac{\partial \tau^*}{\partial T} = \left(\frac{-3\tau_{p0}}{2T_c}\right)\left(\frac{T_c}{T}\right)^{1/2}\left\{1-\left(\frac{T}{T_c}\right)^{1/2}\right\}^2. \tag{5-30}$$

Thus, the strain-rate sensitivity due to the impurities for the F-F is given by substitution of equation (5-30) into equation (5-29) as follows

$$\frac{\partial \tau^*}{\partial \ln \dot{\varepsilon}} = \left(\frac{3\tau_{p0}T}{2T_c}\right)\left(\frac{T_c}{T}\right)^{1/2}\left\{1-\left(\frac{T}{T_c}\right)^{1/2}\right\}^2 \Big/ \alpha, \tag{5-31}$$

where τ_{p0} and T_c for the F-F is given in tables 5-1 and 5-2, and is represented as a solid line in figure 5-4 (a) for KCl:Sr^{2+} (0.050 mol.% in the melt) and (b) for KCl:Sr^{2+} (0.035 mol.% in the melt). The open circles, which correspond to the $(\Delta\tau'/\Delta\ln\dot{\varepsilon})_p$ for the specimen, are approximated to the solid line. Consequently, the F-F seems to be a suitable model for the interaction between a dislocation and the impurity in the specimens.

Substituting of the differential equation of the following equation (see sec. 4.3 of chap. 4) with respect to the temperature into equation (5-29), the strain-rate sensitivity due to the impurities for the Fleischer's model can be obtained.

$$\left(\frac{\tau_{p1}}{\tau_{p0}}\right)^{1/2} = 1-\left(\frac{T}{T_c}\right)^{1/2}. \tag{5-32}$$

Namely,

$$\frac{\partial \tau^*}{\partial \ln \dot{\varepsilon}} = \left(\frac{\tau_{p0}}{T_c}\right)\left\{\left(\frac{T_c}{T}\right)^{1/2} - 1\right\}\frac{T}{\alpha}, \tag{5-33}$$

where τ_{p0} is 14.5 MPa and T_c is 227 K for KCl:Sr^{2+} (0.050 mol.% in the melt). The values of τ_{p0} and T_c for the Fleischer's model is small in contrast to those for the F-F. In addition, the strain-rate sensitivity due to the impurities obtained from equation (5-33) is represented as a dashed line in figure 5-4 (a). Although the difference between the strain-rate sensitivity due to the impurities for the F-F and for the Fleischer's model is not almost observed above approximately 100 K, that for the F-F is large as against that for the Fleischer's model below 100 K for the specimen as shown in figure 5-4 (a). If the $(\Delta\tau'/\Delta\ln\dot{\varepsilon})_p$ below 100 K could be measured for the specimens, the applicability of the Friedel relation would be distinct on the strain-rate sensitivity due to impurities versus the temperature curve.

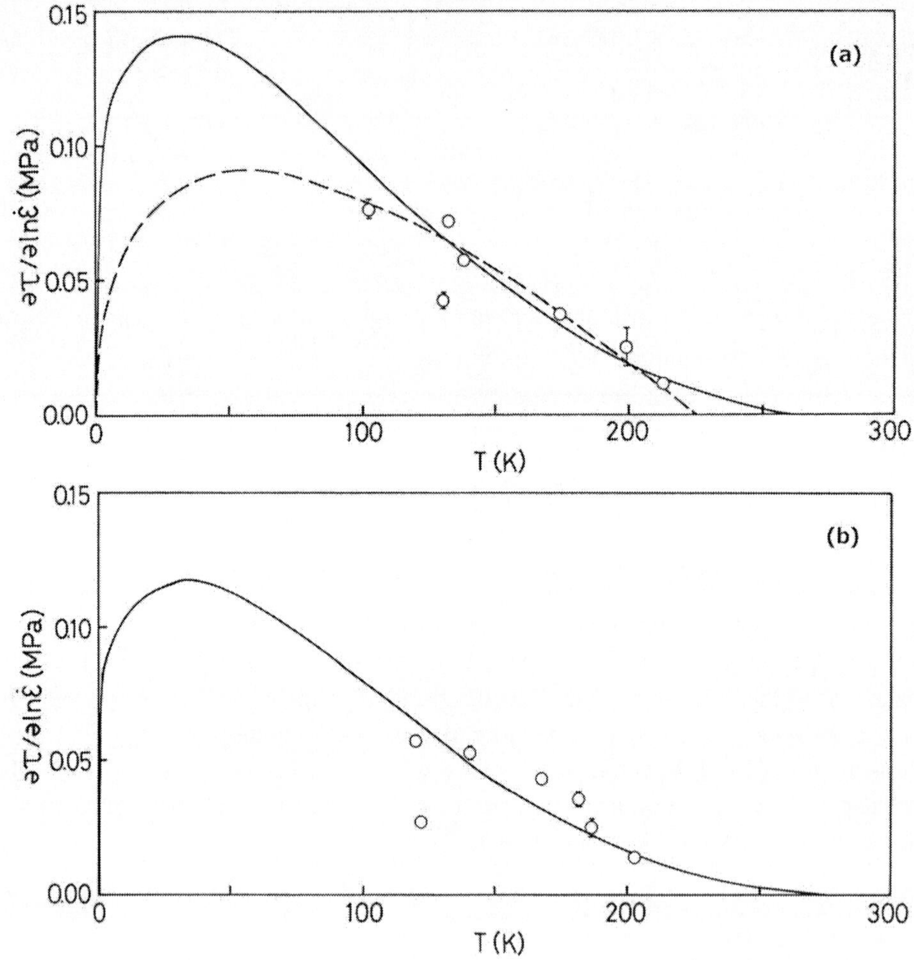

Figure 5-4. Relationship between the strain-rate sensitivity due to the impurities and temperature for KCl:Sr^{2+}: (a) 0.050 mol.%, (b) 0.035 mol.%. Solid line corresponds to the dependence of temperature and strain-rate sensitivity due to the impurities for the F-F and dashed line that for the Fleischer's model. (○) : $(\Delta\tau'/\Delta\ln\dot{\varepsilon})_p$ for the specimens.

5.6. The Enthalpy and the Gibbs Free Energy of Activation for the Break-Away of a Dislocation from the Various Divalent Impurities

The activation enthalpy for the interaction between a dislocation and the impurity is calculated from the substitutional equation of equation (5-30) in equation (5-27), namely

$$\Delta H = -kT^2\left(\frac{\Delta \ln \dot{\varepsilon}}{\Delta \tau'}\right)_p\left(\frac{-3\tau_{p0}}{2T_c}\right)\left(\frac{T_c}{T}\right)^{1/2}\left\{1-\left(\frac{T}{T_c}\right)^{1/2}\right\}^2. \tag{5-34}$$

The values of the activation enthalpy derived from equation (5-34) for the F-F are plotted as a function of temperature in figure 5-5 (b) for $KCl:Ca^{2+}$, (c) for $KCl:Sr^{2+}$, and (d) for $KCl:Ba^{2+}$.

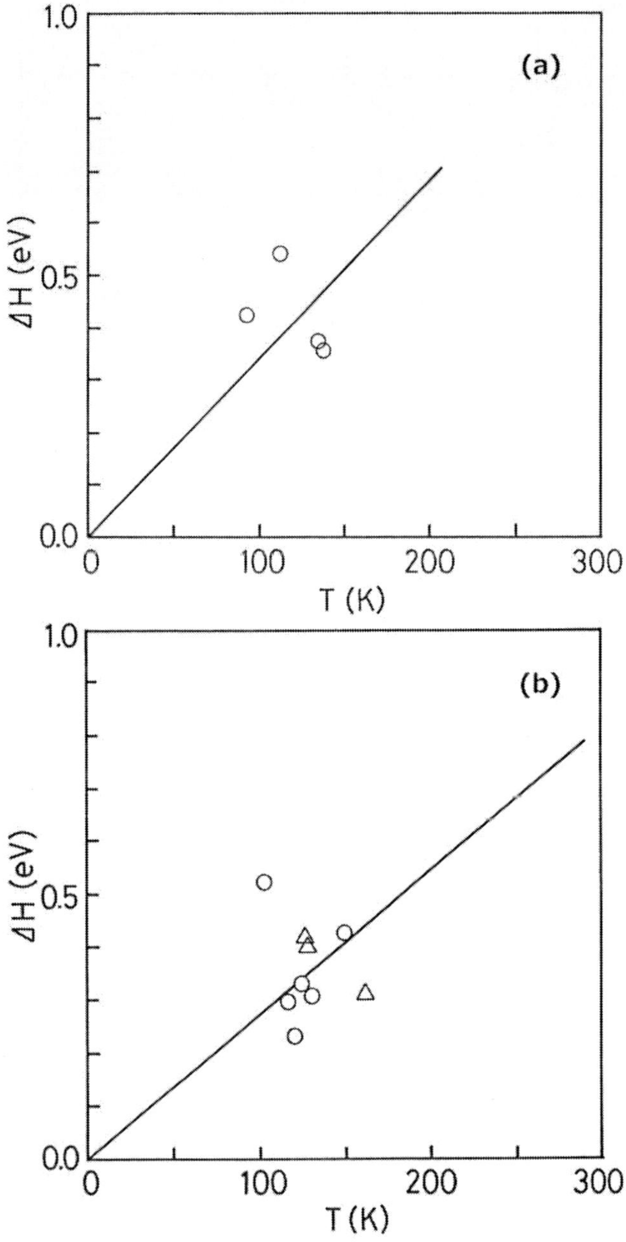

Figure 5-5. (Continued).

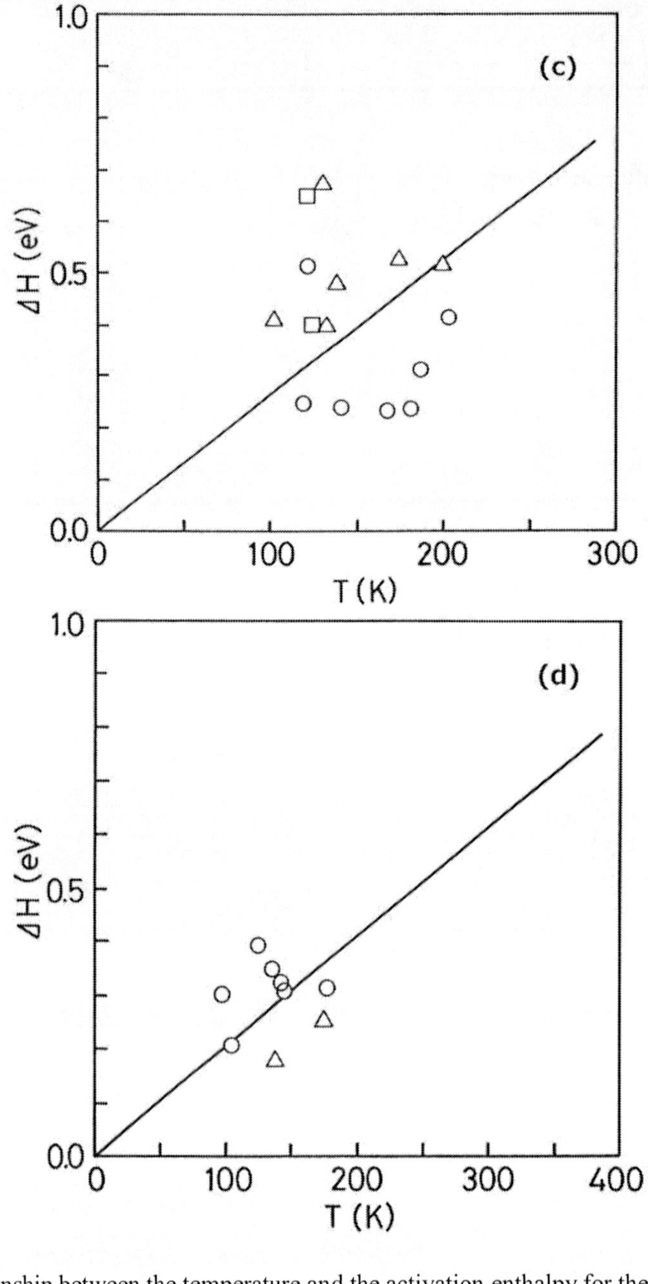

Figure 5-5. Relationship between the temperature and the activation enthalpy for the interaction between a dislocation and the impurity for (a) KCl:Mg^{2+} (○) 0.035 mol.% (b) KCl:Ca^{2+} (○) 0.035 mol.% and (Δ) 0.065 mol.% and (c) KCl:Sr^{2+} (○) 0.035 mol.%, (Δ) 0.050 mol.%, and (□) 0.065 mol.% and (d) KCl:Ba^{2+} (○) 0.050 mol.% and (Δ) 0.065 mol.%.

On the other hand, ΔH for the Fleischer's model is given by equation (5-27), in which $(\partial\tau^* / \partial T)_{\dot{\varepsilon}}$ for the Fleischer's model can be used from the differentiation of equation (5-32) as follows

$$\frac{\partial \tau^*}{\partial T} = \left(\frac{\tau_{p0}}{T_c}\right)\left\{1 - \left(\frac{T}{T_c}\right)^{-1/2}\right\}. \tag{5-35}$$

That is to say, ΔH for the Fleischer's model can be calculated by

$$\Delta H = -kT^2\left(\frac{\Delta \ln \dot{\varepsilon}}{\Delta \tau'}\right)_p\left(\frac{\tau_{p0}}{T_c}\right)\left\{1 - \left(\frac{T}{T_c}\right)^{-1/2}\right\}. \tag{5-36}$$

The calculations by using equation (5-36) is shown in figure 5-5 (a) for KCl:Mg^{2+}. It is seen from figures 5-5 (a) to (d) that the relation between the temperature and the activation enthalpy is satisfied by equation (5-28). The $\Delta H(T_c)$ obtained from figures 5-5 (a) to (d), which corresponds to the activation enthalpy for break-away of the dislocation from the impurity at 0 K, is given in table 5-4. The values of $\Delta H(T_c)$ for the F-F are almost the same for KCl: Ca^{2+}, Sr^{2+} or Ba^{2+} except KCl:Mg^{2+}.

Table 5-4. Values of $\Delta H(T_c)$, ΔG_0 and $\Delta\varepsilon$

Single crystal	$\Delta H(T_c)$ (eV)	ΔG_0 (eV)	$\Delta\varepsilon$
KCl:Mg^{2+}	0.65	—	—
KCl:Ca^{2+}	0.76	0.29	0.20
KCl:Sr^{2+}	0.79	0.39	0.27
KCl:Ba^{2+}	0.75	0.48	0.33

The Gibbs free energy for overcoming the impurity by a dislocation at 0 K is obtained from the following way. Differentiating the combining equations (5-20) and (5-21) with respect to the shear stress gives

$$\frac{\partial \ln \dot{\varepsilon}}{\partial \tau^*} = \left(\frac{2\Delta G_0}{3kT\tau_0}\right)\left(\frac{\tau_0^*}{\tau^*}\right)^{2/3}\left\{1 - \left(\frac{\tau^*}{\tau_0^*}\right)^{1/3}\right\} + \frac{\partial \ln \dot{\varepsilon}_0}{\partial \tau^*}. \tag{5-37}$$

Further substituting of equation (5-22) in equation (5-37) gives

$$\frac{\partial \ln \dot{\varepsilon}}{\partial \tau^*} = \left(\frac{2\Delta G_0}{3kT\tau_{p0}}\right)\left\{1 - \left(\frac{T}{T_c}\right)^{1/2}\right\}^{-2}\left(\frac{T}{T_c}\right)^{1/2} + \frac{\partial \ln \dot{\varepsilon}_0}{\partial \tau^*}, \tag{5-38}$$

where τ^*_0 is replaced by τ_{p0}. Figures 5-6 (a) to (c) show the result of calculations of equation (5-38) for the F-F. The open symbols correspond to the $(\Delta \ln \dot{\varepsilon} / \Delta \tau')_p$ for the specimens. From the slope of straight line, the Gibbs free energy can be obtained as given in table 5-4. The ΔG_0 for KCl:Mg^{2+} could not be obtained here. This will be examined in the subsequent sections 6-7 and 6-8 of chapter 6. ΔG_0 tends to increase with increasing the divalent cationic size in table 5-4. Therefore the tetragonality ($\Delta\varepsilon$) produced around the divalent ion-positive ion vacancy pair is expected to become larger from the Fleischer's model ($F_0 = \mu\Delta\varepsilon b^2/3.86$) [1, 14] as suggested in section 4.5 of chapter 4 when the divalent cationic size increasingly approaches to the K^+ ionic size. The $\Delta\varepsilon$ for KCl: Ca^{2+}, Sr^{2+} or Ba^{2+} is estimated below. The force acted on the dislocation at 0 K is expressed as [1, 13]

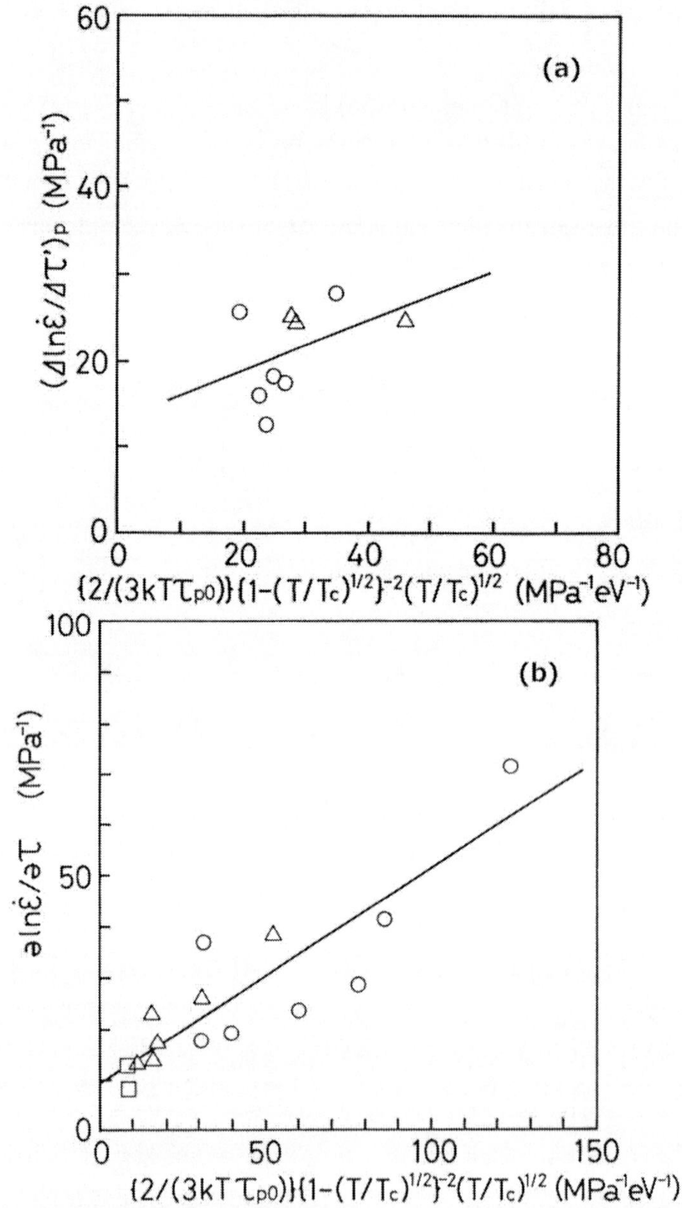

Figure 5-6. Linear plots of equation (5-38) for (a) KCl:Ca^{2+} (○) 0.035 mol.% and (Δ) 0.065 mol.%, (b) KCl:Sr^{2+} (○) 0.035 mol.%, (Δ) 0.050 mol.%, and (□) 0.065 mol.% and (c) KCl:Ba^{2+} (○) 0.050 mol.% and (Δ) 0.065 mol.%.

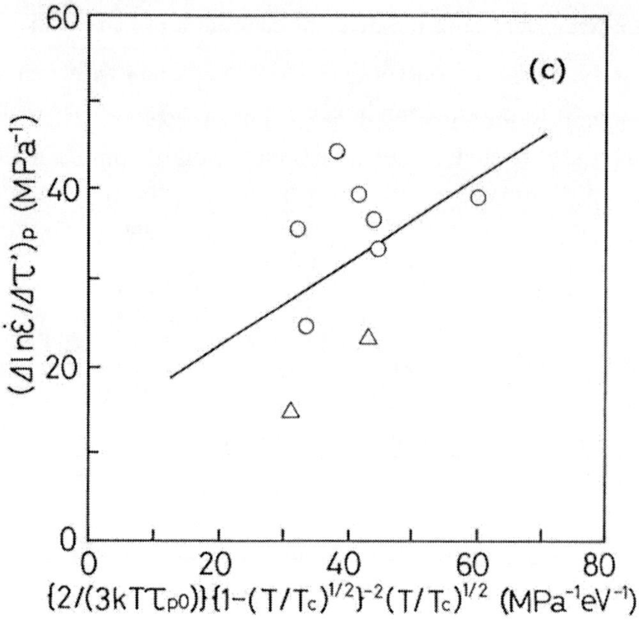

Figure 5-6. (Continued)

$$F_0 = bL_0\mu\Delta\varepsilon c^{1/2}/3.3\,. \qquad (5\text{-}39)$$

The $\Delta\varepsilon$ is calculated from the substitutional equation of equation (5-2) in equation (5-39), namely

$$\Delta\varepsilon = 3.81\Delta G_0/(b^3\mu)\ (\Delta G_0 = F_0 b). \qquad (5\text{-}40)$$

The values of $\Delta\varepsilon$, which is obtained by substituting the value of ΔG_0 in equation (5-40), is given in table 5-4 for the each specimen.

CONCLUSION

On the basis of the $(L_0/L)^2$ versus ϕ_0 curves of the Friedel and the Fleischer relations, the Friedel relation, as well as the Fleischer relation, is considered to be suitable for the weak obstacles which impede the mobile dislocation above ϕ_0 of about 150 degrees. Then, L is approximately over twice the length of L_0. From the values of ϕ_0 for the specimens, the Friedel relation between the effective stress and the average length of dislocation segments seems to be suitable for the interaction between a dislocation and the impurity in KCl: Ca^{2+}, Sr^{2+} or Ba^{2+} single crystals. However, it is seemed that the Friedel relation is not appropriate for KCl:Mg^{2+}.

If the interaction between a dislocation and the impurity can be approximated to the F-F, T_c tends to increase with increasing the divalent cationic size. The values of $\Delta H(T_c)$ for the F-F are approximately the same, but ΔG_0 tends to be larger with increasing the divalent cationic size with regard to the three kinds of specimens except KCl:Mg^{2+}. In addition, $\Delta\varepsilon$ becomes larger when the divalent cationic size increasingly approaches that of the K^+ ion from the small divalent cationic size side. Consequently, a difference in divalent cationic size has an influential factor of the thermally activated flow parameters: T_c, the Gibbs free energy for overcoming the impurity by a dislocation, and $\Delta\varepsilon$.

REFERENCES

[1] Fleischer, R. L. *J. Appl. Phys.* 1962, *33*, 3504.
[2] Friedel, J. *Dislocations;* Pergamon Press; Oxford, 1964; p 224.
[3] Foreman, A. J. E.; Makin, M. J. *Phil. Mag.* 1966, *14*, 911.
[4] Sprackling, M. T. *The Plastic Deformation of Simple Ionic Crystals*, (edited by Alper, A. M.; Margrave, J. L.; Nowick, A. S.) Academic Press; London, 1976; p 188.
[5] Fleischer, R. L.; Hibbard Jr., W. R. *The Relation between Structure and Mechanical Properties of Metals, National Physical Laboratory*, Her Majesty's Stationary Office; London, 1963; Vol. 1, pp 261-297.
[6] Sumino, K. *Jpn. Inst. Met.* 1972, *11*, 31 (in Japanese).
[7] Sumino, K. *Plasticity of Crystals,* Maruzen; Tokyo, 1977; pp 365-374 (in Japanese).
[8] Hart, S. *Brit. J. Appl. Phys.* (*J. Phys. D*) ser. 2, 1968, *1*, 1285.
[9] Conrad, H. *J. Metals* 1964, *16*, 582.
[10] Conrad, H. *J. Phys.* 1967, *45*, 581.
[11] Dey, B. N.; Tyson, W. R. *Phys. Status Solidi* (a) 1972, *9*, 215.
[12] Kataoka, T.; Uematsu, T.; Yamada, T. *Jpn. J. Appl. Phys.* 1978, *17*, 271.
[13] Lloyd, D. J.; Tangri, K. *Phil. Mag.* 1972, *26*, 665.
[14] Fleischer, R. L. *Acta Metall.* 1962, *10*, 835.

Chapter 6

APPLICABILITY OF THE FRIEDEL RELATION TO THE INTERACTION BETWEEN A DISLOCATION AND THE IMPURITY IN KCL:MG^{2+} SINGLE CRYSTALS

6.1. INTRODUCTION

It is well known that the Friedel relation [1] between the effective stress and the average length of dislocation segments is exact for most weak obstacles to dislocation motion at low concentration of the obstacles. It is investigated in this chapter 6 whether the Friedel relation is valid for the interaction between a dislocation and the divelent impurity in KCl:Mg^{2+} (0.035 mol.% in the melt) single crystal, which is expected to cause the "rapid hardening" described in chapter 1. This is carried out from examining various relationships (e.g., effective stress versus temperature, strain-rate sensitivity due to the impurities versus temperature, effective stress versus activation volume, dislocation velocity-effective stress exponent versus temperature, activation enthalpy versus temperature, etc.). The force-distance relation, which represents the interaction between a dislocation and an impurity, is investigated with respect to the two models: one is the Fleischer's model [2] and the other the Fleischer's model taking account of the Friedel relation. The latter is hereafter termed the F-F. Shape of the force-distance curve for short-range obstacle to dislocation motion means the model overcoming the obstacle by a dislocation. This is explained in the chapter 1.

In addition, the activation enthalpy and the Gibbs free energy for overcoming the impurity by a dislocation are obtained here.

6.2. RELATION BETWEEN EFFECTIVE SHEAR STRESS AND TEMPERATURE

The relation between the effective shear stress (τ_{p1}) due to the impurities and temperature is considered to reveal the force-distance relation between a dislocation and an impurity (see sec. 4.3 of chap. 4). The τ_{p1} is considered to represent the effective stress due to only one type of the impurities which lie on the dislocation when the dislocation moves forward with the help of oscillation (see sec. 2.7 of chap. 2). τ_{p1} depends on temperature and on the type and density of impurities (see sec. 2.5 and 2.6 of chap. 2). The relative formula of τ_{p1} and temperature for the Fleischer's model is

$$\left(\frac{\tau_{p1}}{\tau_{p0}}\right)^{1/2} = 1 - \left(\frac{T}{T_c}\right)^{1/2} \tag{6-1}$$

and that for the F-F is expressed in the form :

$$\left(\frac{\tau_{p1}}{\tau_{p0}}\right)^{1/3} = 1 - \left(\frac{T}{T_c}\right)^{1/2}, \tag{6-2}$$

where τ_{p0} is the effective shear stress due to the impurities without thermal activation and T_c is the critical temperature T, at which τ_{p1} is zero, and is independent of an impurity concentration (see sec. 2.5 of chap. 2). τ_{p0} is the value of τ_{p1} at 0 K. The results of equations (6-1) and (6-2) are shown in figures 6-1 (a) and (b) for KCl:Mg^{2+}. Therefore, the interaction between a dislocation and the impurity for the specimen is approximated by the F-F in figure 6-1 (a) and by the Fleischer's model in figure 6-1 (b). The values of T_c and τ_{p0}, which are obtained from figures 6-1 (a) and (b), are given in table 6-1. The two values for the Fleischer's model are small in comparison with those for the F-F. The separation of various models between a dislocation and an impurity has been carried out through the variation of the thermal component of the yield stress with temperature [3]. However, the difference in the linear relationship of effective stress and temperature for the two models cannot be observed within the temperature as shown in figures 6-1 (a) and (b). Ono [4] also concluded as follows. It is impossible to select one of several models on the basis of an experimentally obtained relationship of stress and temperature.

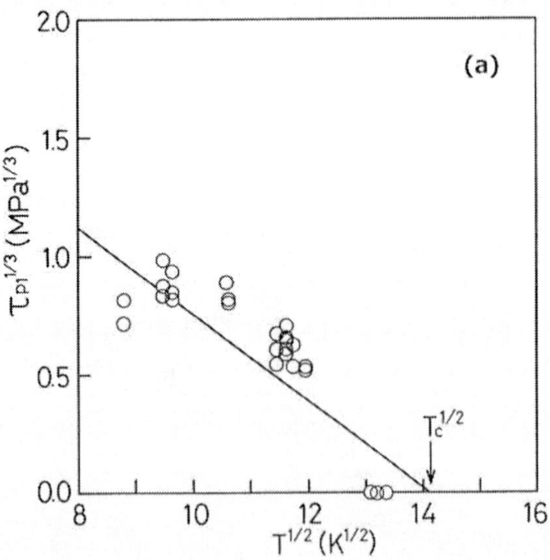

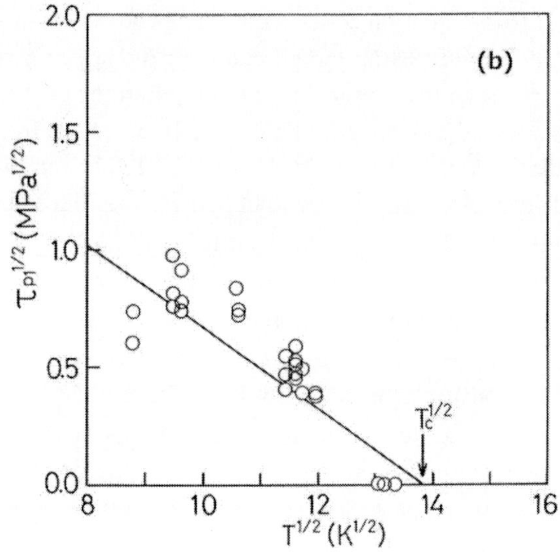

Figure 6-1. Linear plots of the effective shear stress and the temperature for KCl:Mg^{2+} (0.035 mol.% in the melt) at the two models: (a) the F-F and (b) the Fleischer's model.

Table 6-1. Values of τ_{p0} and T_c for KCl:Mg^{2+} (0.035 mol.% in the melt) at the two models

Model	τ_{p0} (MPa)	T_c (K)
Fleischer's model	5.64	191
F-F	17.91	199

6.3. Dependence of the Strain-Rate Sensitivity Due to the Impurities on the Temperature

When the force-distance relation, which represents the interaction between a dislocation and an impurity, can be approximated by the Fleischer's model, the strain-rate sensitivity due to the impurities is represented as

$$\frac{\partial \tau^*}{\partial \ln \dot{\varepsilon}} = \left(\frac{\tau_{p0}}{T_c}\right)\left\{\left(\frac{T_c}{T}\right)^{1/2} - 1\right\}\frac{T}{\alpha}. \tag{6-3}$$

On the other hand, that for the F-F is expressed by

$$\frac{\partial \tau^*}{\partial \ln \dot{\varepsilon}} = \left(\frac{3\tau_{p0}T}{2T_c}\right)\left(\frac{T_c}{T}\right)^{1/2}\left\{1 - \left(\frac{T}{T_c}\right)^{1/2}\right\}^2 \Big/ \alpha. \tag{6-4}$$

The process of leading equations (6-3) and (6-4) was described in the section 5.5 of chapter 5. The values of τ_{p0} and T_c for equation (6-3) can be derived from equation (6-1) and those for equation (6-4) from equation (6-2). Figure 6-2 shows the dependence of the strain-rate sensitivity due to the impurities on temperature for the specimen. The relative curve of the strain-rate sensitivity due to the impurities and temperature for the Fleischer's model and that for the F-F are represented as a solid and a dashed lines, respectively. The two curves are derived from the calculations of equations (6-1) and (6-2). The strain-rate sensitivity due to the impurities for the F-F is obviously larger than that for the Fleischer's model below about 100 K as shown in figure 6-2. The open circles correspond to $(\Delta\tau'/\Delta\ln\dot{\varepsilon})_p$ for the specimen, which are considered to be the strain-rate sensitivity due to impurities (see sec. 4.4 of chap. 4). The $(\Delta\tau'/\Delta\ln\dot{\varepsilon})_p$ is represented by the difference between the strain-rate sensitivities at the first plateau place and at the second one on the relative curve of strain- rate sensitivity and stress decrement as is demonstrated in figure 6-3. Figure 6-3 shows the typical variation of strain-rate sensitivity with $\Delta\tau$ for the specimen. The $\Delta\tau$, which is an axis of abscissa in figure 6-3, is the stress drop due to superposition of the ultrasonic oscillatory stress. The solid line in figure 6-4 shows the relative curve of strain-rate sensitivity and stress decrement for KCl:Mg^{2+} at 113 K. The value of $\Delta\tau'/\Delta\ln\dot{\varepsilon}$ at which $\Delta\tau$ is 0.23 MPa is obviously larger than that at which $\Delta\tau$ is 0 MPa as can be seen from open triangles in figure 6-4. Therefore, the strain-rate sensitivity at the first plateau region is assumed to be in the middle of those at which $\Delta\tau$ is 0 and 0.23 MPa. That is, the $(\Delta\tau'/\Delta\ln\dot{\varepsilon})_p$ in figure 6-4 corresponds to the strain-rate sensitivity due to the impurities for the specimen at 113 K. Figure 6-5 shows the curve for KCl:Mg^{2+} at 143 K. It is the curve at the smallest temperature at which τ_{p1} does not appear. Then, the $(\Delta\tau'/\Delta\ln\dot{\varepsilon})_p$ is assumed to be the difference between the strain-rate sensitivities at which the curve intersects the ordinate and at plateau region, since τ_{p1} first becomes zero at 143 K within the accuracy. Therefore, the $(\Delta\tau'/\Delta\ln\dot{\varepsilon})_p$ at this temperature would be 0.0215 MPa from figure 6-5.

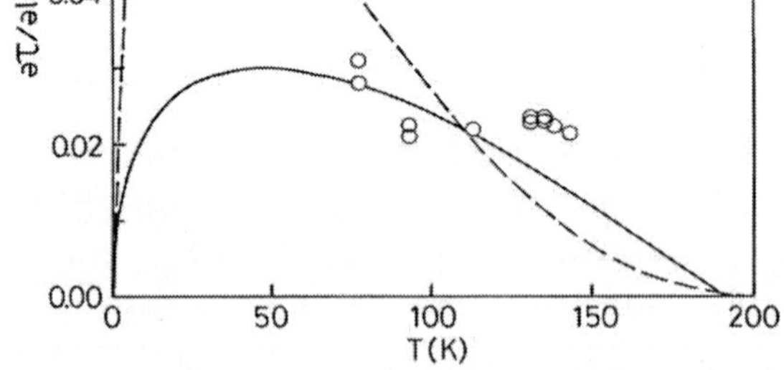

Figure 6-2. Relationship between the strain-rate sensitivity due to the impurities and temperature for KCl:Mg^{2+} (0.035 mol.% in the melt). Solid line corresponds to the dependence of the strain-rate sensitivity on temperature for the Fleischer's model and dashed line that for the F-F. (○) : $(\Delta\tau'/\Delta\ln\dot{\varepsilon})_p$ for the specimen.

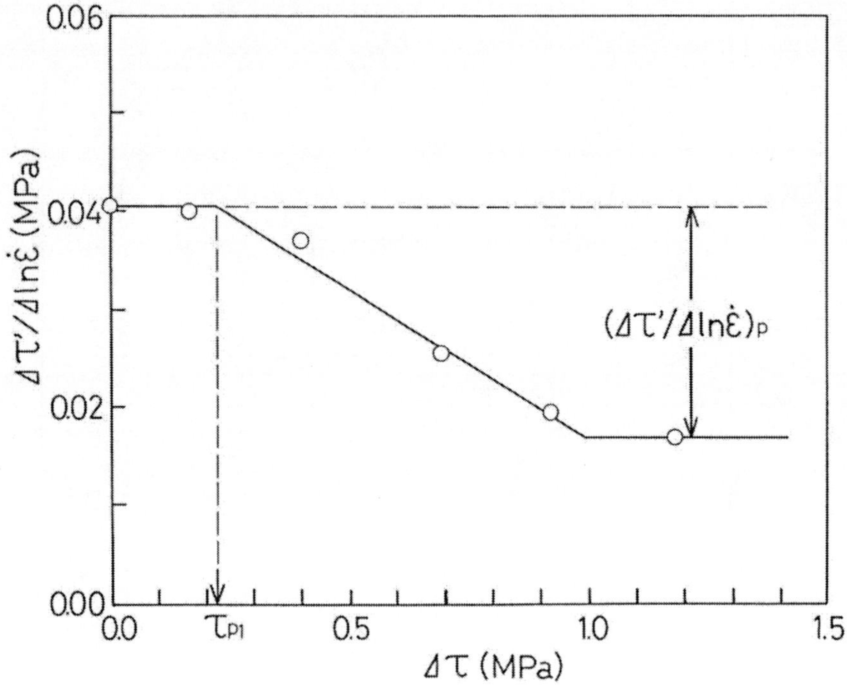

Figure 6-3. Relationship between the strain-rate sensitivity and the stress decrement for KCl:Mg^{2+} (0.035 mol.% in the melt) at 131 K and ε =10%.

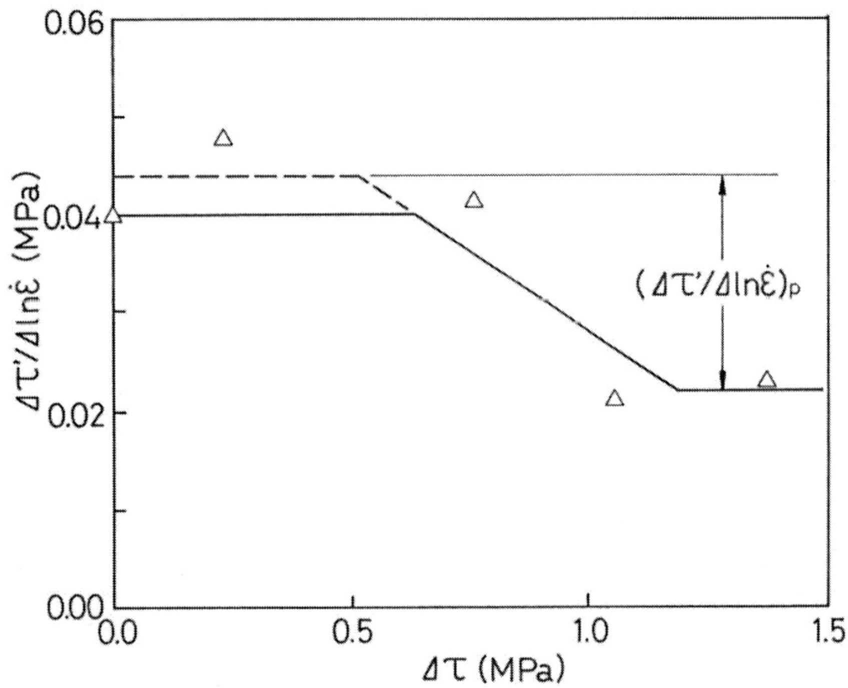

Figure 6-4. Relationship between the strain-rate sensitivity and the stress decrement for KCl:Mg^{2+} (0.035 mol.% in the melt) at 113 K and ε =9%.

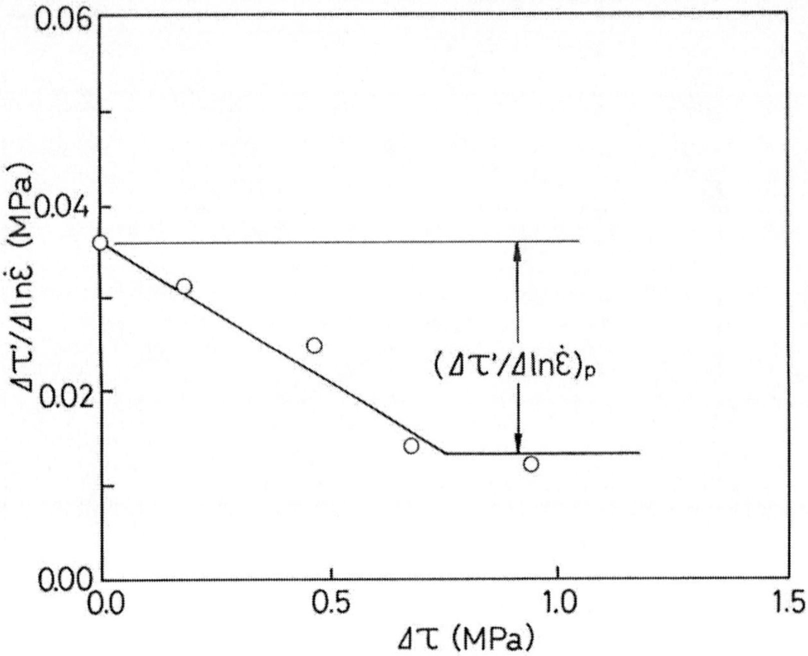

Figure 6-5. Relationship between the strain-rate sensitivity and the stress decrement for KCl:Mg^{2+} (0.035 mol.% in the melt) at 143 K and ε =10%.

As can be seen from figure 6-2, the open circles are approximated to the solid line rather than the dashed one within the present temperatures. That is, the Fleischer's model seems to be more suitable than the F-F. This means that the Friedel relation is not appropriate to the interaction between a dislocation and the impurity in the specimen. If the $(\Delta\tau'/\Delta\ln\dot{\varepsilon})_p$ can be obtained below about 100 K, it will be more distinctly determined whether the Friedel relation is appropriate for the specimen.

6.4. Relation between the Effective Shear Stress and the Activation Volume for the Overcoming of the Impurities by a Dislocation

The measurement of activation volume and its variation with strain (flow stress) have been investigated in order to identify the rate-controlling mechanism to dislocation motion [5-11]. The consideration of the activation volume as dependent on the effective stress is useful in examining the force-distance relation between a dislocation and an impurity [12]. At a given concentration of impurity activation volume varies with effective stress according to the shape of the force-distance curve [13]. The activation volume, υ^*, is given by

$$\upsilon^* = kT\left(\partial \ln \dot{\varepsilon}/\partial \tau^*\right)_T, \tag{6-5}$$

where k is the Boltzmann's constant. The υ^* for the Fleischer's model is expressed by substitution of equation (6-3) into equation (6-5) as follows

$$\upsilon^* = \alpha k\left(\frac{T_c}{\tau_{p0}}\right)\left\{\left(\frac{T_c}{T}\right)^{1/2} - 1\right\}^{-1}. \qquad (6\text{-}6)$$

Similarly from equations (6-4) and (6-5), that for the F-F is expressed by

$$\upsilon^* = \left(\frac{2\alpha kT_c}{3\tau_{p0}}\right)\left(\frac{T}{T_c}\right)^{1/2}\left\{1-\left(\frac{T}{T_c}\right)^{1/2}\right\}^{-2}. \qquad (6\text{-}7)$$

The results of calculations for υ^* at the two models are shown in figure 6-6 for the specimen. The τ_{p1} for the Fleischer's model of ordinate in the figure is obtained from equation (6-1) by the following equation

$$\tau_{p1} = \tau_{p0}\left\{1-\left(\frac{T}{T_c}\right)^{1/2}\right\}^{2} \qquad (6\text{-}8)$$

and that for the F-F from equation (6-2) by

$$\tau_{p1} = \tau_{p0}\left\{1-\left(\frac{T}{T_c}\right)^{1/2}\right\}^{3}. \qquad (6\text{-}9)$$

A solid and a dashed lines represent the relative curve of υ^* and τ_{p1} for the Fleischer's model and that for the F-F, respectively. The open circles in figure 6-6 correspond to the dependence of υ^* calculated from $(\Delta\tau'/\Delta\ln\dot{\varepsilon})_p$ on τ_{p1} for the specimen. That is, the υ^* is given by

$$\upsilon^* = kT(\Delta\ln\dot{\varepsilon}/\Delta\tau')_p. \qquad (6\text{-}10)$$

No difference between the two curves can be almost observed above the υ^* of about 1×10^{-20} cm^3. Below 1×10^{-20} cm^3, however, τ_{p1} for the F-F increases more markedly with decreasing υ^* (temperature) than that for the Fleischer's model. As can be seen from figure 6-6, it is impossible to determine which curve is appropriate for the open circles within the present temperatures. If the $(\Delta\tau'/\Delta\ln\dot{\varepsilon})_p$ can be obtained at still lower temperature, it may be distinguished whether the Friedel relation is appropriate for the specimen from the figure. However, the Peierls mechanism can be adapted to the deformation of alkali halide crystals, such as LiF, NaCl, NaBr and KCl, at the low temperature [14-16], where a dislocation moves forward by the viscous motion. It has been reported for KCl-4 mol% KBr single crystals that the mobile dislocations are impeded by the Peierls barriers in addition to the solute ions below about several tens K [9]. Therefore, there is difficulty in obtaining τ_{p1} and $(\Delta\tau'/\Delta\ln\dot{\varepsilon})_p$ from the relative curve of strain-rate sensitivity and stress decrement for the specimen at the low temperature region at which the Peierls barriers additionally control the dislocation motion. This is because the $\Delta\tau'/\Delta\ln\dot{\varepsilon}$ versus $\Delta\tau$ curve seems to reflect the dislocation motion on the slip plane containing two kinds of obstacles: impurities and forest dislocations, as mentioned in section 2.7 of chapter 2. On the basis of υ^* calculated from $(\Delta\tau'/\Delta\ln\dot{\varepsilon})_p$, namely, from equation (6-10), it is difficult to determine the applicability of the Friedel relation for the specimen.

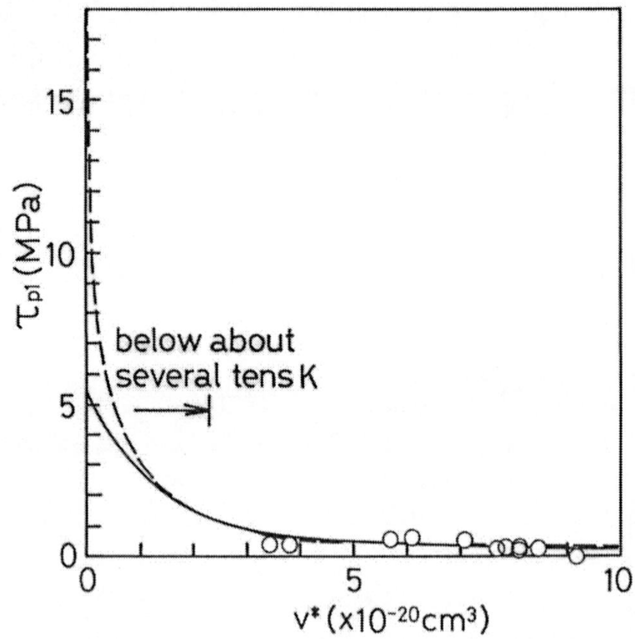

Figure 6-6. Relationship between the effective shear stress and the activation volume for the interaction between a dislocation and the impurities in KCl:Mg^{2+} (0.035 mol.% in the melt). Solid line corresponds to the dependence of the activation volume on the effective shear stress for the Fleischer's model and dashed line that for the F-F. (○) : $kT(\Delta\ln\dot{\varepsilon}/\Delta\tau')_p$ for the specimen.

6.5. CONSIDERATION WITH $m_{\varepsilon=0}$

This study is carried out by the use of dislocation velocity-effective stress exponent expressed by [17]

$$\upsilon = A\tau^{*m^*}, \tag{6-11}$$

where υ is the average velocity of dislocation, A and m^* are constants for a given material and temperature, and τ^* is the effective shear stress due to the impurities. Equation (6-11) is regarded as an empirical relationship [18, 19]; nevertheless, m^* is useful for investigating the behavior of thermally activated dislocation motion in various materials (e.g., NaCl contained Mg^{2+}, Ca^{2+}, Sr^{2+}, and Ba^{2+} [20], body-centered cubic metals [21], Nb [22], LiF:Mg^{2+} [23], and binary iron-base alloys contained Co, Cr, Al, Si, Ni and Mn, respectively [24]). From m^*, rate-controlling mechanism has been examined [25-27].

If the dislocation velocity is controlled by the thermally activated overcoming of the short-range stress field around the I-V dipoles and the dislocation moves forward a distance, L, the dislocation velocity is given by [28]

$$\upsilon = \nu L \exp\left(\frac{-\Delta G}{kT}\right), \tag{6-12}$$

where ν is the frequency of vibration of a dislocation segment of length (L), ΔG is the change in Gibbs free energy of activation for the dislocation motion, and kT has the usual meaning. The ΔG in equation (6-12) is expressed for the Fleischer's model [2] taking account of the Friedel relation [1] as follows (see sec. 5.3 of chap.5)

$$\Delta G = F_0 b \left\{1 - \left(\frac{\tau^*}{\tau_0^*}\right)^{1/3}\right\}^2, \tag{6-13}$$

where F_0 is the force acted on the dislocation at the temperature of 0 K, b is the magnitude of the Burgers vector, and τ^*_0 the effective shear stress due to the impurities at 0 K. Substituting equation (6-13) in equation (6-12) gives

$$\upsilon = \nu L \exp\left[- F_0 b \left\{1 - \left(\frac{\tau^*}{\tau_0^*}\right)^{1/3}\right\}^2 \Big/ (kT)\right]. \tag{6-14}$$

Natural logarithmic differentiation of equation (6-11) with respect to the shear stress gives

$$m^* = \frac{\Delta \ln \upsilon}{\Delta \ln \tau^*} = \tau^* \left(\frac{\Delta \ln \upsilon}{\Delta \tau^*}\right). \tag{6-15}$$

The $(\Delta \ln \upsilon / \Delta \tau^*)$ in equation (6-15) can be obtained from equation (6-14) as follows

$$\frac{\Delta \ln \upsilon}{\Delta \tau^*} = \frac{2F_0 b}{3kT\tau_0^*}\left\{\left(\frac{\tau^*}{\tau_0^*}\right)^{-2/3} - \left(\frac{\tau^*}{\tau_0^*}\right)^{-1/3}\right\}. \tag{6-16}$$

Substituting equation (6-16) in equation (6-15), the m^* for the F-F is expressed by

$$m^{*} = 2F_0 b\left\{\left(\frac{\tau^{*}}{\tau_0^{*}}\right)^{1/3} - \left(\frac{\tau^{*}}{\tau_0^{*}}\right)^{2/3}\right\}\Big/(3kT). \tag{6-17}$$

Furthermore, substituting the equation (6-2) between effective shear stress and temperature, where τ_{p1} and τ_{p0} are replaced by τ^{*} and τ^{*}_0 respectively, into equation (6-17), the m^{*} for the F-F can be evaluated as

$$m^{*} = 2F_0 b\left\{\left(\frac{T_c}{T}\right)^{1/2} - 1\right\}\Big/(3kT_c), \tag{6-18}$$

The F_0 in equation (6-18) is determined by (see equation (5-18) in sec. 5.3 of chap. 5, where τ^{*}_0 is replaced by τ_{p0})

$$F_0 = \left(2L_0^2 E\right)^{1/3}\left(b\tau_{p0}\right)^{2/3}, \tag{6-19}$$

where L_0 is the average spacing of impurities on the slip plane and E the line tension of the dislocations. The line tension of the dislocations is calculated by μb^2. The shear modulus (μ) for the [110] direction is assumed to be 1.01×10^{10} Pa at 0 K [27]. The average spacing of impurities on the slip plane has been given by [2, 29]

$$L_0 = \frac{b}{\left(\frac{4c}{3}\right)^{1/2}}, \tag{6-20}$$

where c is the concentration of impurities and is determined to be 7.9 ppm by dielectric loss measurement.

On the other hand, the ΔG in equation (6-12) is expressed for the Fleischer's model as follows [2]

$$\Delta G = F_0 b\left\{1 - \left(\frac{\tau^{*}}{\tau_0^{*}}\right)^{1/2}\right\}^{2}. \tag{6-21}$$

Combination of equations (6-12) and (6-21) yields

$$\upsilon = \nu L \exp\left[-F_0 b\left\{1 - \left(\frac{\tau^{*}}{\tau_0^{*}}\right)^{1/2}\right\}^{2}\Big/(kT)\right]. \tag{6-22}$$

Natural logarithmic differentiation of equation (6-22) with respect to $\ln\tau^{*}$ gives [2]

$$m^* = F_0 b\left\{\left(\frac{\tau^*}{\tau_0^*}\right)^{1/2} - \left(\frac{\tau^*}{\tau_0^*}\right)\right\} \Big/ (kT). \tag{6-23}$$

Substituting equation (6-1), where τ_{p1} and τ_{p0} are replaced by τ^* and τ^*_0 respectively, into equation (6-23), we find

$$m^* = F_0 b\left\{\left(\frac{T_c}{T}\right)^{1/2} - 1\right\} \Big/ (kT_c), \tag{6-24}$$

where F_0 is determined by

$$F_0 = \tau_{p0} Lb. \tag{6-25}$$

The m^* of equation (6-24) corresponds to that for the Fleischer's model. Temperature versus m^* curves calculated from equations (6-18) and (6-24) are represented by a dashed and a solid lines in figure 6-7. Then, the values of T_c and τ_{p0} for the F-F and the Fleischer's model are tabulated in table 6-1. The two curves show that m^* increase with decreasing temperature. The m^* value for the F-F is greater than that for the Fleischer's model at a given temperature. The difference between m^* values for the two models also increases with decreasing temperature, which ranges 3.9 to 28.6 at the temperature of 190 to 90 K.

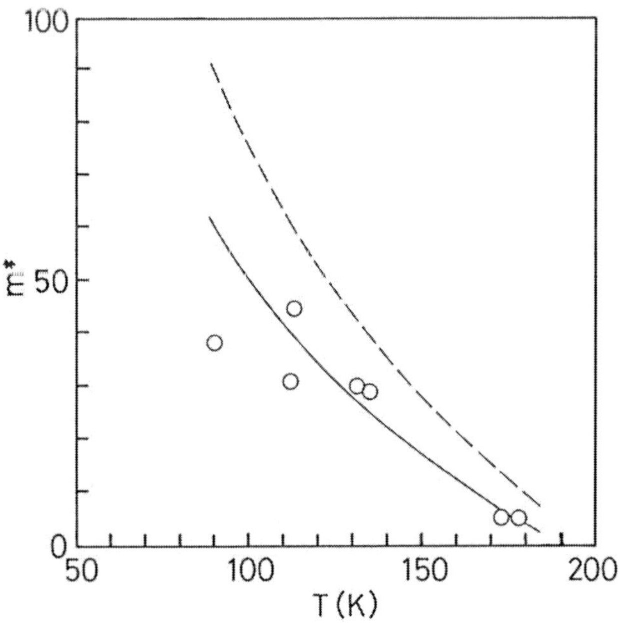

Figure 6-7. Relationship between the temperature and the dislocation velocity-effective stress exponent for KCl:Mg^{2+} (0.035 mol.% in the melt) at the two models: (- - -) the F-F and (—) the Fleischer's model. Open circles represent the $m_{\varepsilon=0}$ for the specimen.

According to the theory by Nix *et al.* [30], the value of m^* is identical with m when τ_i is considerably small in comparison with τ_a. m is a dislocation velocity-applied stress exponent. τ_i and τ_a are internal stresses and applied stress, respectively. The work hardening in both stage I (easy glide region) and stage II (linear hardening region) of plastic deformation region was largely due to internal stress components in LiF [31] and NaCl [20, 31, 32] single crystals. Then, the effective stress due to a small amount of impurity was almost constant independently of strain during the plastic deformation. Therefore, the value of m^* seems to almost agree with $m_{\varepsilon=0}$ (m value at zero strain) for a given material and temperature. Furthermore, the influence of internal stresses on dislocation velocity in zinc could be neglected at the applied shear stress below about 0.01 MPa [33]. This also implies that the value of m^* $(=\Delta \ln \upsilon / \Delta \ln \tau^*)$ corresponds to m $(=\Delta \ln \upsilon / \Delta \ln(\tau^* + \tau_i))$ at extremely small strain and a given temperature. Johnston and Stein [34] have proposed that m^* can be determined by extrapolating m to zero strain for LiF single crystal. Evans and Pratt [32] showed that m^* compares well with m at zero strain in CaF_2 single crystals. In addition, the values of m^* and $m_{\varepsilon=0}$ for NaCl single crystals at 244 K were 13 ± 2 and 16 ± 2, respectively [32]. The $m_{\varepsilon=0}$ was independent of the strain-rate increments: $\dot{\varepsilon}_1/\dot{\varepsilon}_2$ =2, 4, 10, 40, 100, 400, and 1000 [32]. $m_{\varepsilon=0}$ for alkali halide was considered to be m^*, since the thermal components of flow stress are independent of mobile dislocation density at low concentration of impurities and near zero strain, where the thermally activated overcoming of impurities by a dislocation is dominant [35]. Deformation behavior has been investigated with $m_{\varepsilon=0}$ so far [19, 32, 36-38]. Using $m_{\varepsilon=0}$, reasonable value of effective stress and hardening mechanism were found for zone-refined iron [36] and potassium contained Na^+ (25 ppm) [37] during the deformation.

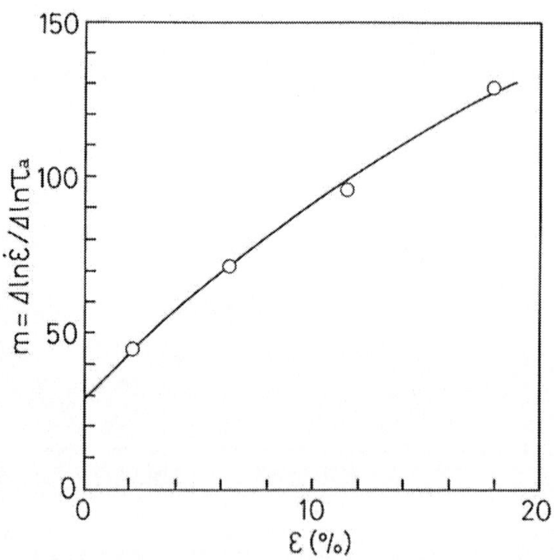

Figure 6-8. Variation of m with shear strain for $KCl{:}Mg^{2+}$ (0.035 mol.% in the melt) at 135 K.

Figure 6-8 shows m versus shear strain curve for KCl:Mg^{2+} (0.035 mol.%). The m values are obtained here by strain-rate cycling tests without applying the oscillation. Through the m value extrapolated to zero strain in figure 6-8, therefore, m^* seems to be about 29 at 135 K on the assumption that internal stresses and mobile dislocation density of the specimen are unchanged during the change in strain rate, because m^* is defined as [34]

$$\left(\Delta \ln \dot{\varepsilon}/\Delta \ln \tau^*\right)_T = m^* + \left(\Delta \ln \rho/\Delta \ln \tau^*\right)_T, \tag{6-26}$$

where ρ is the density of mobile dislocations. The $(\Delta \ln \dot{\varepsilon}/\Delta \ln \tau^*)_T$ in this study is given by $(1.609/\Delta \ln \tau^*)$. Open circles in figure 6-7 correspond to the dependence of temperature and the $m_{\varepsilon=0}$ for KCl:Mg^{2+} (0.035 mol.%). The open circles tend to increase with decreasing temperature and approach to solid line. This suggests that the interaction between a dislocation and the impurity in the specimen can be approximated to the Fleischer's model rather than to the F-F. That is, the F-F is considered to be unsuitable for the specimen. Consequently, it may be deduced that the Friedel relation is not appropriate for the interaction between a dislocation and the impurity in the specimen within the temperature range. This estimation with dislocation velocity-effective stress exponent was carried out in the narrow range of temperature; nevertheless, this method is clearer than any other judgment here in chapter 6.

6.6. Relation between Temperature and the Activation Enthalpy for the Overcoming of the Impurities by a Dislocation

When a dislocation overcomes the impurity with the aid of thermal fluctuation, activation enthalpy (ΔH) for the Fleischer's model is given by (see sec. 5.6 of chap. 5)

$$\Delta H = -kT^2 \left(\frac{\Delta \ln \dot{\varepsilon}}{\Delta \tau'}\right)_p \left(\frac{\tau_{p0}}{T_c}\right) \left\{1 - \left(\frac{T}{T_c}\right)^{-1/2}\right\}. \tag{6-27}$$

On the other hand, ΔH for the F-F is calculated from (see sec. 5.6 of chap. 5)

$$\Delta H = -kT^2 \left(\frac{\Delta \ln \dot{\varepsilon}}{\Delta \tau'}\right)_p \left(\frac{-3\tau_{p0}}{2T_c}\right) \left(\frac{T_c}{T}\right)^{1/2} \left\{1 - \left(\frac{T}{T_c}\right)^{1/2}\right\}^2. \tag{6-28}$$

Assuming that the changes in entropy are neglected, ΔH is in proportion to the temperature [39, 40]. That is, ΔH is expressed by

$$\Delta H = \alpha kT. \tag{6-29}$$

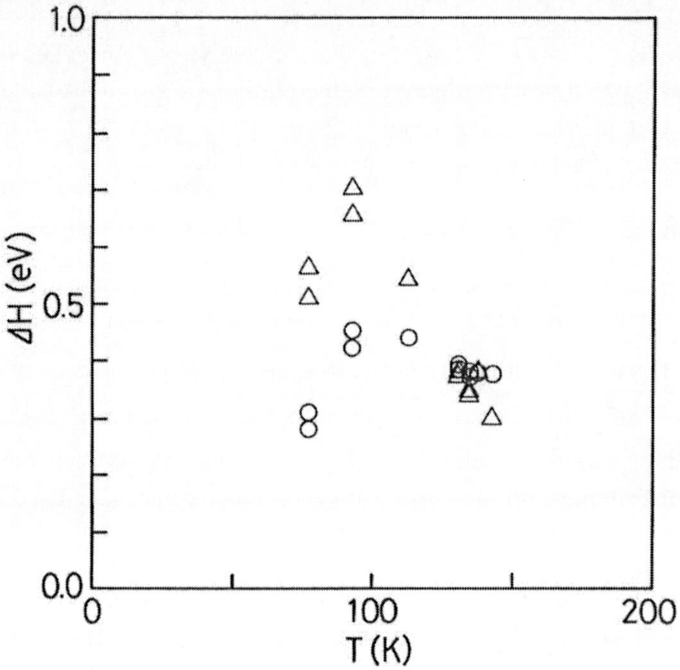

Figure 6-9. Relationship between the temperature and the activation enthalpy for the interaction between a dislocation and the impurity in KCl:Mg^{2+} at the two models: (○) the Fleischer's model and (Δ) the F-F.

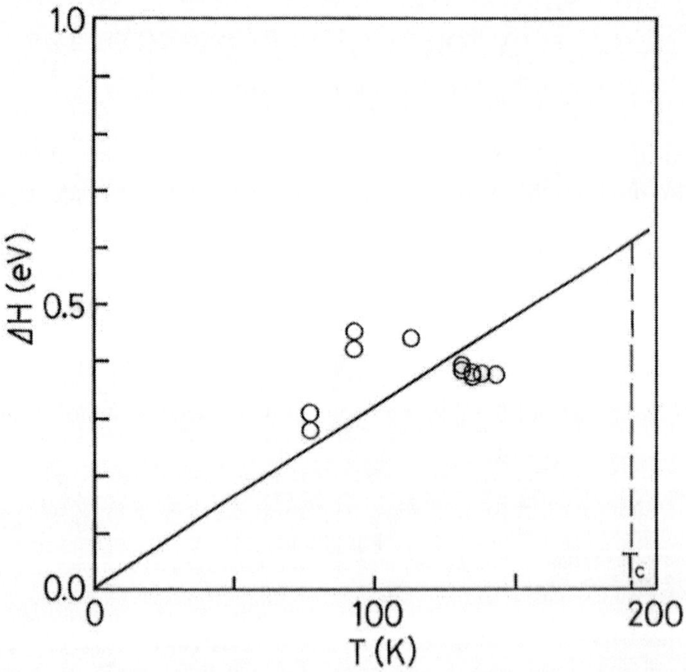

Figure 6-10. Proportional relationship between the temperature and the activation enthalpy for the interaction between a dislocation and the impurity in KCl:Mg^{2+} at the Fleischer's model.

With respect to the two models (the Fleischer's model and the F-F), the proportional relationships of temperature and the activation enthalpy is investigated below for the specimen. The results of equations (6-27) and (6-28) are represented by open circles and triangles in figure 6-9. The open circles appear to be nearly proportional to the temperature, compared with the open triangles. The $\Delta H(T)$ for the Fleischer's model seems to be satisfied with equation (6-29) within the temperature range. Then, it can be found that the Fleischer's model is suitable for the specimen rather than the F-F from figure 6-9. Therefore, the Friedel relation seems not to be appropriate to the interaction between a dislocation and the impurity in the specimen.

Figure 6-10 shows the proportional relation between temperature and the activation enthalpy at the Fleischer's model for the specimen. The value of $\Delta H(T_c)$ taken from figure 6-10 is 0.61 eV. It corresponds to the activation enthalpy for overcoming of the strain field around the I-V dipole by a dislocation without an applied stress. The value of $\Delta H(T_c)$ at the F-F is 0.73 eV.

6.7. CONSIDERATION WITH $\tau_{P1}(\Delta \mathrm{LN}\,\dot{\varepsilon}/\Delta\tau')_P$

When the force-distance relation between a dislocation and an impurity can be approximated by the F-F, the m^* may be also given by equation (6-17) transformed into

$$m^* = \frac{2\Delta G_0}{3kT}\left\{\left(\frac{\tau_{p1}}{\tau_{p0}}\right)^{1/3} - \left(\frac{\tau_{p1}}{\tau_{p0}}\right)^{2/3}\right\}, \tag{6-30}$$

where ΔG_0 is the Gibbs free energy for the break-away of the dislocation from the impurity in the absence of an applied stress and is obtained from the following equation (see sec. 5.6 of chap. 5)

$$\frac{\partial \ln \dot{\varepsilon}}{\partial \tau^*} = \left(\frac{2\Delta G_0}{3kT\tau_{p0}}\right)\left(\frac{T}{T_c}\right)^{1/2}\left\{1-\left(\frac{T}{T_c}\right)^{1/2}\right\}^{-2} + \frac{\partial \ln \dot{\varepsilon}_0}{\partial \tau^*}. \tag{6-31}$$

The result of calculations of equation (6-31) is shown by open circles in figure 6-11. As for the Fleischer's model, the m^* may be given by the following equation in accordance with equation (6-23), namely,

$$m^* = \frac{\Delta G_0}{kT}\left\{\left(\frac{\tau_{p1}}{\tau_{p0}}\right)^{1/2} - \left(\frac{\tau_{p1}}{\tau_{p0}}\right)\right\}, \tag{6-32}$$

where ΔG_0 is obtained from

$$\frac{\partial \ln \dot{\varepsilon}}{\partial \tau^*} = \left(\frac{\Delta G_0}{kT\tau_{p0}}\right)\left(\frac{T}{T_c}\right)^{1/2}\left\{1-\left(\frac{T}{T_c}\right)^{1/2}\right\}^{-1} + \frac{\partial \ln \dot{\varepsilon}_0}{\partial \tau^*}. \tag{6-33}$$

Open circles in figure 6-12 show the results of equation (6-33). The $\partial \ln \dot{\varepsilon}/\partial \tau^*$ in equations (6-31) and (6-33) corresponds to the $(\Delta \ln \dot{\varepsilon}/\Delta \tau')_p$, which is given by the reciprocal of difference between the strain-rate sensitivities at first plateau place and at second one on the relative curve of strain-rate sensitivity and stress decrement, for the specimen in stage II (linear hardening region) of stress-strain curve. The $(\Delta \tau'/\Delta \ln \dot{\varepsilon})_p$ is considered to be the strain-rate sensitivity due to the impurities (see sec. 4.4 of chap. 4). Accordingly, $\partial \ln \dot{\varepsilon}_0/\partial \tau^*$ in equations (6-31) and (6-33) could be evaluated from $\Delta \ln \dot{\varepsilon}_0/\Delta \tau'$, which is represented by the solid circles in figures 6-11 and 6-12. On the basis of the slope of dashed lines in figures 6-11 and 6-12, the ΔG_0 for the specimen at the F-F was estimated to be within the range of 0.40 to 0.46 eV and that at the Fleischer's model 0.42 to 0.48 eV. The process of leading equation (6-33) and ΔG_0 at the Fleischer's model is described below section 6.8. The m^* values calculated from equations (6-30) and (6-32) are denoted by open squares and triangles in figure 6-13. These open symbols tend to increase slightly with decreasing temperature. Furthermore, m^* may be analyzed in terms of the strain-rate sensitivity versus stress decrement curve due to oscillation as follows

$$m^* = \tau_{p1}\left(\Delta \ln \dot{\varepsilon}/\Delta \tau'\right)_p. \tag{6-34}$$

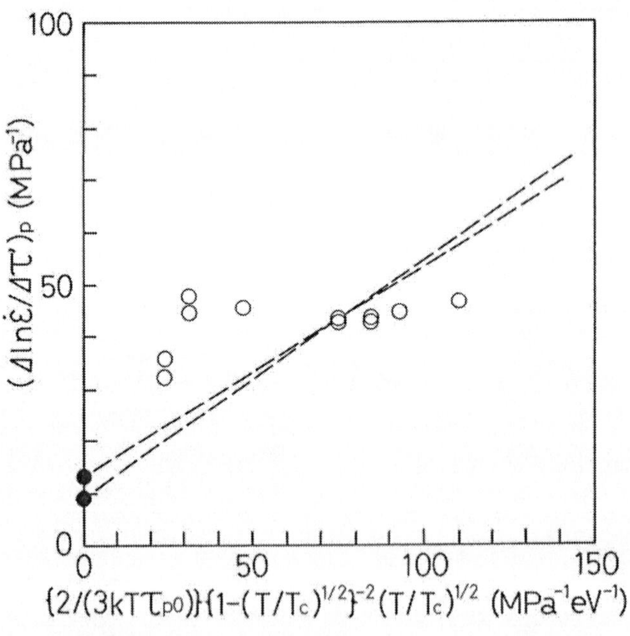

Figure 6-11. Linear plots of equation (6-31) for KCl:Mg^{2+} (0.035 mol.% in the melt). (○) : $(\Delta \ln \dot{\varepsilon}/\Delta \tau')_p$; (●) : $\partial \ln \dot{\varepsilon}_0/\partial \tau^*$.

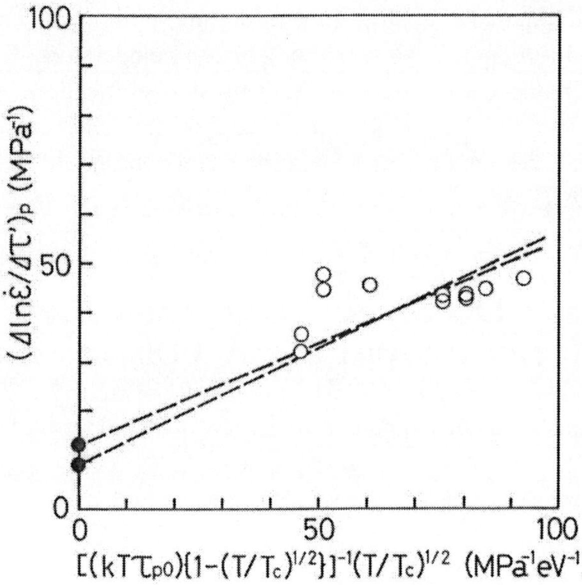

Figure 6-12. Linear plots of equation (6-33) for KCl:Mg^{2+} (0.035 mol.% in the melt). (○) : $(\Delta\ln\dot{\varepsilon}/\Delta\tau')_p$; (●) : $\partial \ln \dot{\varepsilon}_0 / \partial \tau^*$.

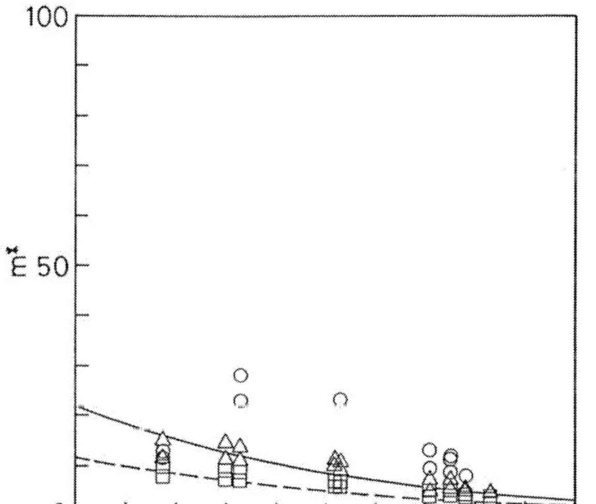

Figure 6-13. Relationship between the temperature and the dislocation velocity-effective stress exponent for KCl:Mg^{2+} (0.035 mol.% in the melt) at the two models: (□) the F-F and (Δ) the Fleischer's model. Open circles represent $\tau_{p1}(\Delta\ln\dot{\varepsilon}/\Delta\tau')_p$ for the specimen.

The τ_{p1} is obtained on the basis of the strain-rate sensitivity versus stress decrement curves due to oscillation, as shown in figure 6-3. This m^* is indicated by open circles in figure 6-13. The values of m^* in equations (6-30), (6-32), and (6-34) are examined on the assumption that equation (6-15) is also valid for the case of applying the ultrasonic oscillatory stress during plastic deformation. As can be seen from figure 6-13, the open circles are closer to m^* for the Fleischer's model in contrast to that for the F-F, though the difference of the two (i.e., open circles and triangles) is great at a given temperature. This means that the interaction between a dislocation and the impurity in the specimen can be fitted to the Fleischer's model rather than to the F-F. Resulting in the same description as preceding section 6.5, it is considered that the Friedel relation is not appropriate for the specimen within the temperature range.

6.8. Gibbs Free Energy for Overcoming of Tetragonal Lattice Distortions by a Dislocation

When a dislocation overcomes the impurity in accordance with the Fleischer's model, the Gibbs free energy of the activation, ΔG, is expressed by [2]

$$\Delta G = \Delta G_0 \left\{ 1 - \left(\frac{\tau^*}{\tau_0^*} \right)^{1/2} \right\}^2, \quad \left(\Delta G_0 = \tau_0^* L b^2 \right) \tag{6-35}$$

where τ^*_0 is the effective shear stress (τ^*) at the temperature of 0 K. From an Arrhenius type equation for the thermally activated deformation rate ($\dot{\varepsilon}$), the Gibbs free energy is also expressed as

$$\Delta G = \alpha k T, \quad \left(\alpha = \ln\left(\dot{\varepsilon}_0 / \dot{\varepsilon} \right) \right) \tag{6-36}$$

where $\dot{\varepsilon}_0$ is a frequency factor. Differentiating the combining equations (6-35) and (6-36) with respect to the shear stress gives

$$\frac{\partial \ln \dot{\varepsilon}}{\partial \tau^*} = \left(\frac{\Delta G_0}{k T \tau_0^*} \right) \left(\frac{\tau_0^*}{\tau^*} \right)^{1/2} \left\{ 1 - \left(\frac{\tau^*}{\tau_0^*} \right)^{1/2} \right\} + \frac{\partial \ln \dot{\varepsilon}_0}{\partial \tau^*}. \tag{6-37}$$

Substituting equation (6-1) in equation (6-37) gives equation (6-33), where τ^*_0 is replaced by τ_{p0}.

The open circles in figure 6-12 show the results of equation (6-33). Only from the data, it is difficult to obtain the ΔG_0 for the specimen on the basis of the slope of a line. One examines the value of $(\Delta \ln \dot{\varepsilon} / \Delta \tau')_p$ at which the line intersects the ordinate in figure 6-12. It corresponds to the $\partial \ln \dot{\varepsilon}_0 / \partial \tau^*$ in equation (6-33). The $\dot{\varepsilon}_0$ is usually expressed by [27, 28]

$$\dot{\varepsilon}_0 = \rho b^2 \nu_D \left(L_0 / L \right)^2, \tag{6-38}$$

where ρ is the density of mobile dislocations, ν_D is the Debye frequency, and L_0 is the average spacing of impurities on the slip plane. Differentiating the natural logarithmic equation of equation (6-38) with respect to the effective shear stress gives

$$\frac{\partial \ln \dot{\varepsilon}_0}{\partial \tau^*} = \frac{\partial \ln \rho}{\partial \tau^*} - 2\left(\frac{\partial \ln L}{\partial \tau^*}\right). \tag{6-39}$$

The $\partial \ln L/\partial \tau^*$ is zero in the Fleischer's model. Therefore, the $\partial \ln \dot{\varepsilon}_0/\partial \tau^*$ in equation (6-33) for the Fleischer's model can be expressed from equation (6-39) by

$$\frac{\partial \ln \dot{\varepsilon}_0}{\partial \tau^*} = \frac{\Delta \ln \dot{\varepsilon}_0}{\Delta \tau'} = \frac{\Delta \ln \rho}{\Delta \tau'}. \tag{6-40}$$

The $\Delta\rho/\Delta\tau'$ is obtained in the subsequent section 11.3 of chapter 11 for KCl doped with Ca^{2+} (0.035 and 0.065 mol.% in the melt), Sr^{2+} (0.035, 0.050, and 0.065 mol.% in the melt) or Ba^{2+} (0.050 and 0.065 mol.% in the melt) in stage II. However, no difference of $\Delta\rho/\Delta\tau'$ among the three kinds of specimens can be discriminated because of scattering $(\Delta \ln \dot{\varepsilon}/\Delta\tau')_p$. The results of $\Delta \ln \rho/\Delta\tau'$ are tabulated in table 6-2. The value of $\Delta \ln \rho/\Delta\tau'$ ranges from 8.94 to 13.06 MPa^{-1} and may not be influenced by the size and the concentration of divalent impurity. Assuming that the magnitude of the $\Delta \ln \rho/\Delta\tau'$ for $KCl{:}Mg^{2+}$ is around those for the KCl doped with Ca^{2+}, Sr^{2+} or Ba^{2+}, the $\partial \ln \dot{\varepsilon}_0/\partial \tau^*$ in equation (6-33) is represented by solid circles in figure 6-12. Then, the ΔG_0, which is obtained from the slope of dashed lines, seems to be between 0.42 and 0.48 eV for the specimen.

Table 6-2. Values of c and $\Delta \ln \rho/\Delta\tau'$ at 0 K

Single crystal (mol.%)	c (ppm)[a]	$\Delta \ln \rho/\Delta\tau'$ (MPa^{-1})
$KCl{:}Ca^{2+}$ (0.035)	43.1	13.06
(0.065)	43.5	13.05
$KCl{:}Sr^{2+}$ (0.035)	55.2	8.94
(0.050)	98.3	8.97
(0.065)	121.8	8.98
$KCl{:}Ba^{2+}$ (0.050)	9.2	12.38
(0.065)	28.3	12.41

a The concentration of divalent impurities is obtained from dielectric loss measurement.

When KCl single crystals are doped with Mg^{2+}, tetragonal lattice distortions are expected to be produced around the I-V dipoles. The tetragonality ($\Delta\varepsilon$) is calculated from the following equation (see sec. 5.6 of chap. 5) :

$$\Delta\varepsilon = 3.81\Delta G_0/(b^3\mu), \tag{6-41}$$

where μ is the shear modulus for the [110] direction at 0 K and is assumed to be 1.01×10^{10} Pa [41]. The value of $\Delta\varepsilon$ for KCl:Mg^{2+} would be within the range of 0.29 to 0.33.

CONCLUSION

It was investigated at 77–178 K whether the Friedel relation is appropriate for the interaction between a dislocation and the impurity in KCl:Mg^{2+} (0.035 mol.% in the melt) single crystals. This was carried out by using various characteristics with respect to the two models: one is the Fleischer's model taking account of the Friedel relation and the other the Fleischer's model. The former model is termed the F-F. As a result, the following was found out.

1. The dependence of strain-rate sensitivity due to the impurities, $(\Delta\tau'/\Delta\ln\dot{\varepsilon})_p$, on temperature for the specimen was appropriate to the Fleischer's model rather than the F-F.
2. The activation enthalpy (ΔH) for the Fleischer's model appeared to be nearly proportional to the temperature in comparison with the F-F.
3. Fitting $m_{\varepsilon=0}$ (the dislocation velocity-applied stress exponent (m) value at zero strain) to the temperature versus dislocation velocity-effective stress exponent (m^*) curves, the $m_{\varepsilon=0}$ values tend to increase with decreasing temperature and further approach to the m^* for the Fleischer's model in contrast to the F-F at the temperature. Similar result was also deduced on the basis of the data analyzed in terms of strain-rate sensitivity versus stress decrement due to oscillation. That is, the m^* values proposed by $\tau_{p1}(\Delta\ln\dot{\varepsilon}/\Delta\tau')_p$ are closer to the m^* for the Fleischer's model, compared with that for the F-F. The discussion in this study is progressed on the assumption that equation (6-15) can be also applied to the case of applying the ultrasonic oscillatory stress during plastic deformation.
4. Assuming that the changes in entropy are neglected, the value of $\Delta H(T_c)$, which corresponds to the activation enthalpy for overcoming of the strain field around the impurity by a dislocation in the absence of an applied stress, for the Fleischer's model was found to be 0.61 eV. In addition, the Gibbs free energy (ΔG_0) concerning the dislocation motion was determined to be between 0.42 and 0.48 eV. Then, the $\Delta\varepsilon$ for the specimen would be within the range of 0.29 to 0.33 on the basis of the value of ΔG_0.

Nos. 1 to 3 of above-mentioned conclusions show that the Fleischer's model is more suitable than the F-F for the interaction between a dislocation and the impurity in KCl:Mg^{2+} single crystal. This leads to the estimation that the Friedel relation between effective stress and average length of the dislocation segments is inappropriate to that in the specimen within the temperature, though it is exact for most weak obstacles to dislocation motion.

REFERENCES

[1] Friedel, J. *Dislocations;* Pergamon Press; Oxford, 1964; p 224.
[2] Fleischer, R. L. *J. Appl. Phys.* 1962, *33*, 3504.
[3] Conrad, H. *Acta Metall.* 1966, *14*, 1631.
[4] Ono, K. *J. Appl. Phys.* 1968, *39*, 1803.
[5] Evans, A. G.; Rawlings, R. D. *Phys. Status Solidi* 1969, *34*, 9.
[6] Lloyd, D. J.; Worthington, P. J.; Embury, J. D. *Phil. Mag.* 1970, *21*, 1147.
[7] Srinivasan, M.; Stoebe, T. G. *J. Mater. Sci.* 1973, *8*, 1567.
[8] Lloyd, D. J.; Chung, D. W.; Chaturvedi, M. C. *Acta Metall.* 1975, *23*, 93.
[9] Kataoka, T.; Uematsu, T.; Yamada, T. *Jpn. J. Appl. Phys.* 1978, *17*, 271.
[10] Saha, G. G.; Mccormick, P. G.; Rama Rao, P. *Mater. Sci. Engng.* 1984, *62*, 197.
[11] Sivakesavam, O.; Saha, G. G.; Srinivasa Raghavan, K. *Mater. Sci. Technol.* 1988, *4*, 578.
[12] Sumino, K. *Jpn. Inst. Met.* 1972, *11*, 135 (in Japanese).
[13] Wielke, B. Czech. *J. Phys. B* 1981, *31*, 142.
[14] Suzuki, T.; Kim, H. *J. Phys. Soc. Japan* 1975, *39*, 1566.
[15] Suzuki, T.; Kim, H. *J. Phys. Soc. Japan* 1976, *40*, 1703.
[16] Suzuki, T. *Oyo Buturi* 1976, *45*, 447 (in Japanese).
[17] Johnston, W. G.; Gilman, J. J. *J. Appl. Phys.* 1959, *30*, 129.
[18] Christian, J. W. *Acta Metall.* 1964, *12*, 99.
[19] Takeuchi, N.; Terada, K.; Yonetani, S. *Zairyo* 1978, *27*, 176 (in Japanese).
[20] Sprackling, M. T. *Phil. Mag.* 1973, *27*, 265.
[21] Anglada, M.; Guiu, F. *Scripta Metall.* 1979, *13*, 103.
[22] Guiu, F.; Anglada, M. *Phil. Mag.* 1982, *46*, 881.
[23] Gorri, J. A.; Paz, A.; Guiu, F. *Phys. Status Solidi* (a) 1984, *82*, 85.
[24] Okazaki, K. *J. Mater. Sci.* 1996, *31*, 1087.
[25] Balasubramanian, N. *Scripta Metall.* 1969, *3*, 21.
[26] Lloyd, D. J.; Worthington, P. J.; Embury, J. D. *Phil. Mag.* 1970, *22*, 1147.
[27] Lloyd, D. J.; Tangri, K. *Phil. Mag.* 1972, *26*, 665.
[28] Sprackling, M. T. *The Plastic Deformation of Simple Ionic Crystals*, (edited by Alper, A. M.; Margrave, J. L.; Nowick, A. S.) Academic Press; London, 1976; p 139.
[29] Sprackling, M. T. *The Plastic Deformation of Simple Ionic Crystals*, (edited by Alper, A. M.; Margrave, J. L.; Nowick, A. S.) Academic Press; London, 1976; p 141.
[30] Nix, W. D.; Coghlan, W. A.; Barrett, C. R. *Mater. Sci. Engng.* 1969, *4*, 98.
[31] Gupta, I.; Li, J. C. M. *Mater. Sci. Engng.* 1970, *6*, 20.
[32] Evans, A. G.; Pratt, P. L. *Phil. Mag.* 1970, *21*, 951.
[33] Adams, K. H. *J. Appl. Phys.* 1968, *39*, 4040.
[34] Johnston, W. G.; Stein, D. F. *Acta Metall.* 1963, *11*, 317.
[35] Imura, T. In *Strength of Crystals*, (edited by Hasiguti, R. R.; Chikazumi, S.) Asakurashoten; Tokyo, 1968; p 28 (in Japanese).
[36] Michalak, J. T. *Acta Metall.* 1965, *13*, 213.
[37] Bernstein, I. M.; LI, J. C. M.; Gensamer, M. *Acta Metall.* 1967, *15*, 801.
[38] Lee, K. S.; Park, W. K. *J. Korean Nucl. Soc.* 1978, *10*, 73.
[39] Conrad, H. *J. Metals* 1964, *16*, 582.

[40] Conrad, H. *J. Phys.* 1967, *45*, 581.
[41] Hart, S. *Brit. J. Appl. Phys.* (*J. Phys. D*) ser. 2, 1968, *1*, 1285.

Chapter 7

INFLUENCE OF STATE OF IMPURITIES ON THE DEFORMATION CHARACTERISTICS

7.1. INTRODUCTION

If the I-V dipoles in $KCl:Sr^{2+}$ single crystals diffuse into the matrix, or they aggregate by heat treatment, various deformation characteristics will be changed.

This is because the state of a small amount of impurities in the crystals strongly influences the resistance to movement of a dislocation as Dryden et al. [1] reported.

In chapter 4, the influence of the heat treatment on the force-distance relation between a dislocation and the impurity was described from the relation between temperature and τ_{p1} (see sec. 4.3 of chap. 4) and the proportionability of $\Delta H(T)$ (see sec. 4.5 of chap. 4) for the two models: triangular force-distance profile and the Fleischer's model.

This chapter 7 concerns additionally the various models (a square, a parabolic, and a triangular force-distance relations between a dislocation and an aggregate and the Fleischer's model taking account of the Friedel relation).

In chapter 7, the change in deformation characteristics (e.g., the critical temperature (T_c) at which the effective stress is zero, the force acted on the dislocation, bending angle of dislocation by the impurity, the temperature dependence of dislocation velocity-effective stress exponent (m^*), the force-distance profile between a dislocation and the impurity, etc.) are continuously described for the various models in more detail.

This is based on the data analyzed from the relative curves of strain-rate sensitivity ($\Delta\tau'/\Delta\ln\dot{\varepsilon}$) and stress decrement ($\Delta\tau$) due to the oscillation.

The specimens used here are two kinds of ones: the quenched and the annealed $KCl:Sr^{2+}$ (0.050 mol.% in the melt) single crystals, which are the same specimens as described in chapter 4.

The agglomerates in the annealed specimen form at least trimers by the heat treatment (see sec. 3.3 of chap. 3).

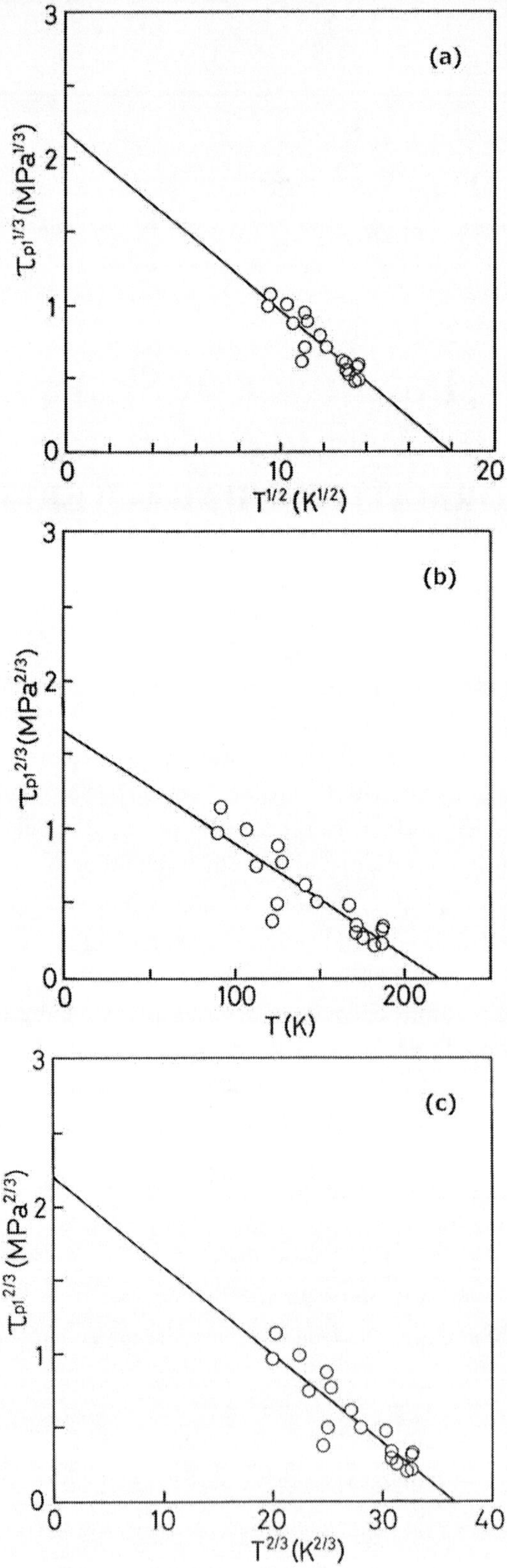
(a)
τp1^1/3 (MPa^1/3)
0
1
2
3
0
10
20
T^1/2 (K^1/2)
(b)
τp1^2/3 (MPa^2/3)
0
1
2
3
0
100
200
T (K)
(c)
τp1^2/3 (MPa^2/3)
0
1
2
3
0
10
20
30
40
T^2/3 (K^2/3)

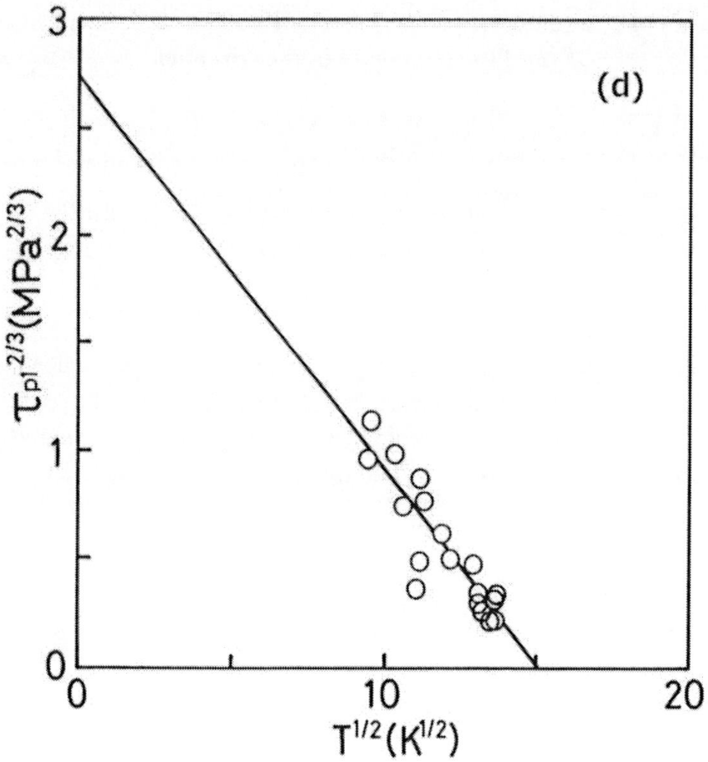

Figure 7-1. Linear plots of the effective shear stress and the temperature for the annealed specimen at various models: (a) the F-F, (b) the SQ, (c) the PA, and (d) the TR.

7.2. Discussion for the Force-Distance Relation between a Dislocation and the Aggregate

7.2.1. Relationship of τ_{P1} and Temperature

Firstly, it is assumed that the interaction between a dislocation and the aggregate for the specimen can be approximated to the Fleischer's model [2] taking account of the Friedel relation [3]. The model is named the F-F here. Then, the relative formula of τ_{p1} and temperature, which will reveal the force-distance relation between a dislocation and an impurity, is given by

$$\left(\frac{\tau_{p1}}{\tau_{p0}}\right)^{1/3} = 1 - \left(\frac{T}{T_c}\right)^{1/2}. \tag{7-1}$$

The result of equation (7-1) is shown in figure 7-1 (a) for the annealed specimen. The values of τ_{p0} and the critical temperature (T_c), at which τ_{p1} is zero, obtained from figure 7-1 (a) are given in table 7-1. The τ_{p0} is obtained by extrapolating the line to 0 K and is considered the effective stress due to the aggregates without thermal activation.

Table 7-1. Values of T_c and τ_{p0} for various force-distance relations between a dislocation and the aggregate in the specimen

The force-distance relation	T_c (K)	τ_{p0} (MPa)
F-F	322	10.30
SQ	220	2.17
PA	225	3.28
TR	228	4.53

Secondly, the relative formula of τ_{p1} and temperature is examined with respect to the three models [4]: a square force-distance relation, a parabolic one and a triangular one, which are hereafter termed the SQ, the PA and the TR, respectively, in the similar way. The three force-distance relations are also taken account of the Friedel relation. That for the SQ is given by

$$\left(\frac{\tau_{p1}}{\tau_{p0}}\right)^{2/3} = 1 - \left(\frac{T}{T_c}\right). \tag{7-2}$$

The PA gives

$$\left(\frac{\tau_{p1}}{\tau_{p0}}\right)^{2/3} = 1 - \left(\frac{T}{T_c}\right)^{2/3}, \tag{7-3}$$

and the TR gives

$$\left(\frac{\tau_{p1}}{\tau_{p0}}\right)^{2/3} = 1 - \left(\frac{T}{T_c}\right)^{1/2}. \tag{7-4}$$

The results of equations (7-2) to (7-4) are shown in figures 7-1 (b) to (c) for the annealed specimen. The slopes of straight lines are determined by the method of least squares.

The separation of various models between a dislocation and an impurity for a particular case has been carried out through the variation of the thermal component of the flow stress with temperature [5]. However, no difference in the linear effective stress-temperature relation for the four models can be observed within the temperature as shown in figures 7-1 (a) to (c). Ono [6] also described about the applied stress-temperature relation as follows: it is impossible to select one of several models on the basis of an experimentally obtained relationship of stress and temperature.

7.2.2. Relative Curve of Temperature and the Strain-Rate Sensitivity due to Aggregates

The relative curve of temperature and the strain-rate sensitivity due to the aggregates for the F-F is represented as a solid line in figure 7-2. The strain-rate sensitivity is calculated from the following equation

$$\frac{\partial \tau^*}{\partial \ln \dot{\varepsilon}} = \left(\frac{3\tau_{p0}T}{2T_c}\right)\left(\frac{T_c}{T}\right)^{1/2}\left\{1-\left(\frac{T}{T_c}\right)^{1/2}\right\}^2 \bigg/ \alpha, \quad (7\text{-}5)$$

where α is an arbitrary constant. The open circles represent the $(\Delta\tau'/\Delta\ln\dot{\varepsilon})_p$ for the annealed specimen, which is given by the difference between strain-rate sensitivity at first plateau region and at second one on the relative curve of strain-rate sensitivity and stress decrement and is the strain-rate sensitivity due to impurities (see sec. 2.1 of chap. 2 and sec. 4.4 of chap. 4). The process of leading equation (7-5) is already described in section 5.5 of chapter 5. The open circles seem to be distant from the solid line below 150 K. Therefore, it is difficult to approximate the interaction between a dislocation and the aggregate in the annealed specimen to the F-F.

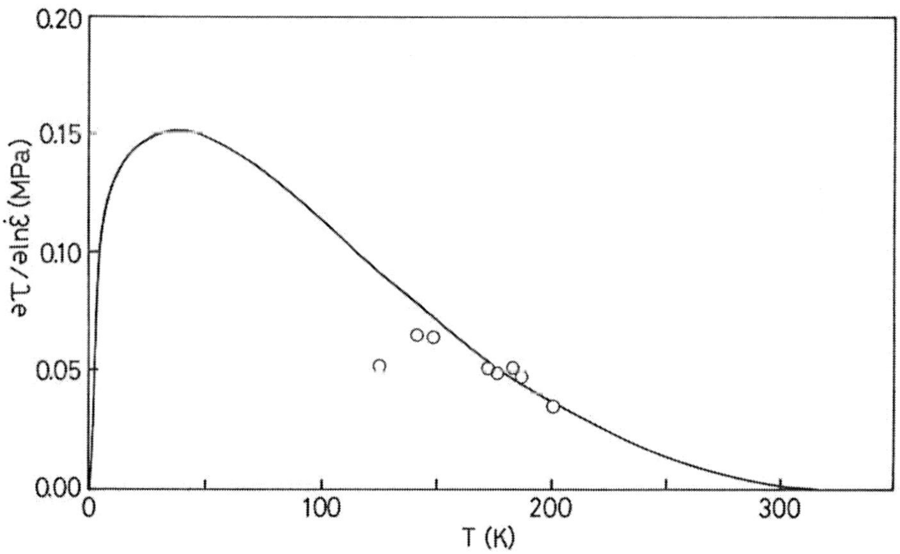

Figure 7-2. Relationship between the strain-rate sensitivity due to the aggregates and temperature for the annealed specimen. Solid line corresponds to the dependence of temperature and the strain-rate sensitivity for the F-F. (○) : $(\Delta\tau'/\Delta\ln\dot{\varepsilon})_p$ for the annealed specimen.

Furthermore, the strain-rate sensitivity due to the aggregates for the SQ is obtained as

$$\frac{\partial \tau^*}{\partial \ln \dot{\varepsilon}} = \left(\frac{3\tau_{p0}}{2T_c}\right)\left\{1-\left(\frac{T}{T_c}\right)\right\}^{1/2} T \bigg/ \alpha \quad (7\text{-}6)$$

That for the PA is expressed by

$$\frac{\partial \tau^*}{\partial \ln \dot{\varepsilon}} = \left(\frac{\tau_{p0}}{T_c}\right)\left\{\left(\frac{T_c}{T}\right)^{2/3} - 1\right\}^{1/2} T \Big/ \alpha \tag{7-7}$$

and also that for the TR is expressed by

$$\frac{\partial \tau^*}{\partial \ln \dot{\varepsilon}} = \left(\frac{3\tau_{p0}}{4T_c}\right)\left\{\left(\frac{T_c}{T}\right) - \left(\frac{T_c}{T}\right)^{1/2}\right\}^{1/2} T \Big/ \alpha . \tag{7-8}$$

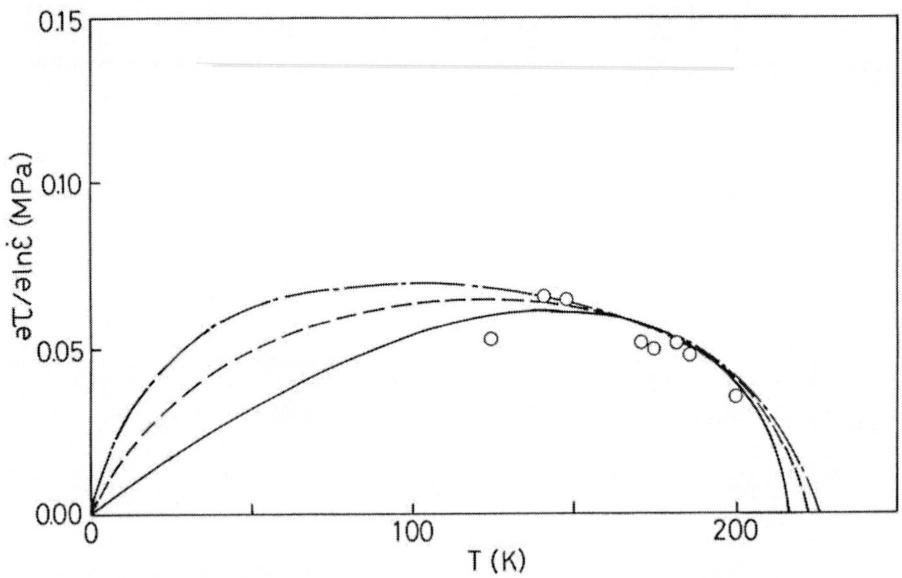

Figure 7-3. Relationship between the strain-rate sensitivity due to the aggregates and temperature for the annealed specimen. Solid line corresponds to the dependence of temperature and the strain-rate sensitivity for the SQ, dashed line does that for the PA, and dot-dash line does that for the TR. (○): $(\Delta\tau'/\Delta\ln\dot{\varepsilon})_p$ for the annealed specimen.

The values of T_c and τ_{p0}, which are obtained from figures 7-1 (b) to (d), are given in table 7-1. The process of leading equations (7-6) to (7-8) is described in the subsequent section 8.7 of chapter 8. Figure 7-3 shows the strain-rate sensitivity due to the aggregates for the specimen. The curves for the SQ, the PA and the TR are derived from the calculations of equations (7-6) to (7-8). The open circles correspond to the $(\Delta\tau'/\Delta\ln\dot{\varepsilon})_p$ for the annealed specimen. As can be seen from figure 7-3, the model between a dislocation and the aggregate in the specimen seems to be nearly equal to the SQ among the three within the temperature. Therefore, the interaction between a dislocation and the aggregate may be approximated to the SQ within the temperature. If the $(\Delta\tau'/\Delta\ln\dot{\varepsilon})_p$ for the specimen can be obtained below about 100 K, the most suitable force-distance relation between a dislocation and the aggregate will be distinctly selected among the three from figure 7-3. However, the values below 100 K could not be unfortunately obtained.

7.2.3. Dislocation Velocity-Effective Stress Exponent

7.2.3.1. Quenched Specimen

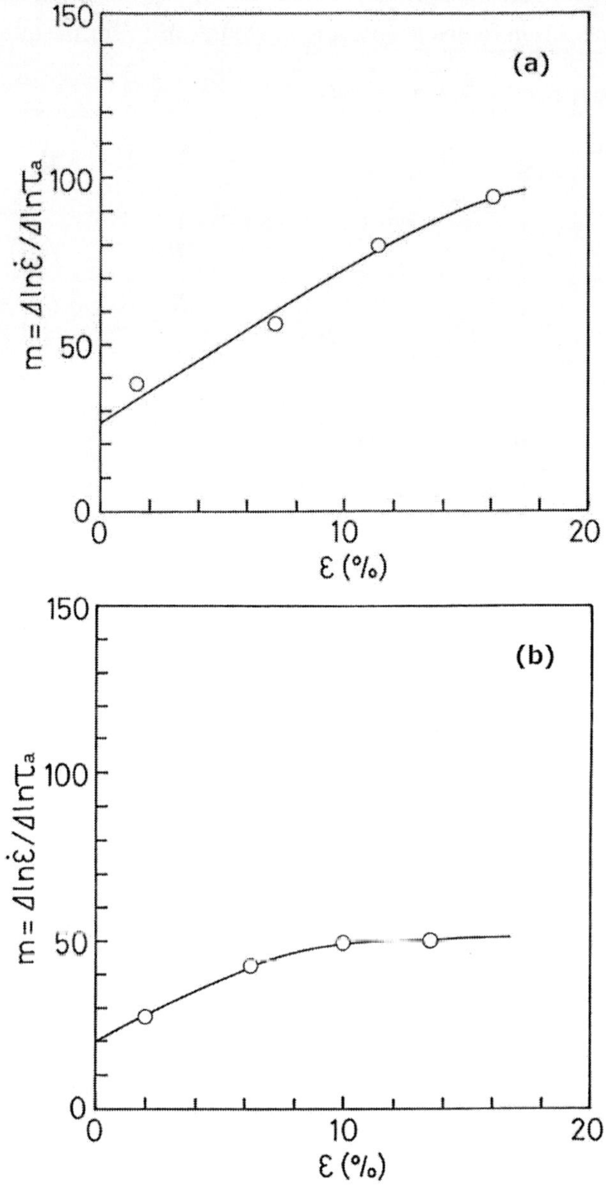

Figure 7-4. Variation of m with shear strain for (a) the quenched specimen at 175 K and (b) the annealed specimen at 125 K.

Deformation behavior has been investigated with the $m_{\varepsilon=0}$ which is obtained by extrapolating a relative curve of strain-rate sensitivity, $\Delta\ln\dot{\varepsilon}/\Delta\ln\tau_a$, and shear strain to zero strain [7-11]. $\Delta\ln\tau_a$ is the logarithmic change in applied shear stress due to the strain-rate cycling test without the oscillation during plastic deformation. The value of $m_{\varepsilon=0}$ for alkali halide was considered to be m^* [12] (i.e., the values of m^* are about 26.5 for the quenched specimen at 175 K and 20.0 for the annealed specimen at 125 K, as can be seen from figures 7-4 (a) and (b)). Using $m_{\varepsilon=0}$, reasonable value of effective stress and hardening mechanism were found for zone-refined iron [8] and potassium contained Na^+ (25 ppm) [9] during the deformation. Open circles in figure 7-5 correspond to the $m_{\varepsilon=0}$ for the quenched specimen.

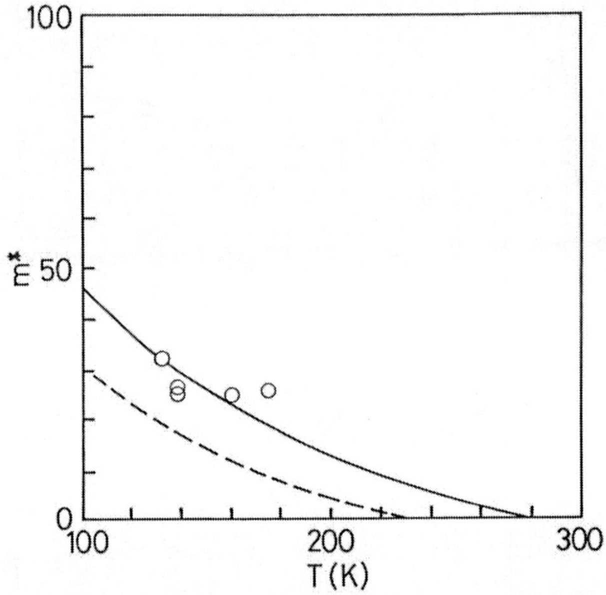

Figure 7-5. Relationship between the temperature and the dislocation velocity-effective stress exponent for the quenched specimen at the two models: (- - -) the Fleischer's model and (—) the F-F. Open circles represent the $m_{\varepsilon=0}$ for the quenched specimen.

The m^* for the Fleischer's model was given by [2] as:

$$m^* = F_0 b\left\{\left(\frac{\tau^*}{\tau_0^*}\right)^{1/2} - \left(\frac{\tau^*}{\tau_0^*}\right)\right\} \Big/ (kT), \tag{7-9}$$

where F_0 is the force acted on the dislocation at 0 K, b is the magnitude of the Burgers vector, τ^*_0 is the effective shear stress τ^* at 0 K, and kT has the usual meaning. The Fleischer's model has been widely used for the interaction between a dislocation and an impurity in ionic crystals doped with divalent cations [2, 13-15]. The temperature dependence of the effective stress was given by [2]

$$\left(\frac{\tau^*}{\tau_0}\right)^{1/2} = 1 - \left(\frac{T}{T_c}\right)^{1/2}. \tag{7-10}$$

Combining equations (7-9) and (7-10), we can evaluate the m^* for the Fleischer's model as follows

$$m^* = F_0 b\left\{\left(\frac{T_c}{T}\right)^{1/2} - 1\right\} \Big/ (kT_c). \tag{7-11}$$

Friedel [1] derived an expression for the average spacing of weak obstacles along the dislocation. The m^* for the Fleischer's model taking account of the Friedel relation, which is abbreviated as F-F in this chapter, can be obtained from the equation:

$$m^* = 2F_0 b\left\{\left(\frac{T_c}{T}\right)^{1/2} - 1\right\} \Big/ (3kT_c). \tag{7-12}$$

Table 7-2. Values of F_0 and T_c for various force-distance relations between a dislocation and the impurity in the two specimens

Single crystal	force-distance relation	F_0 ($\times 10^{-10}$ N)	T_c (K)
Quenched specimen	Fleischer	5.25	227
	F-F	9.20	289
Annealed specimen	SQ	2.85	220
	PA	3.75	225
	TR	4.65	228

The method of leading equation (7-12) is particularized in section 6.5 of chapter 6. The results of equations (7-11) and (7-12), in which F_0 and T_c are given in table 7-2, are represented as a dashed and a solid lines respectively in figure 7-5. The value of F_0 for the Fleischer's model is calculated from

$$F_0 = \tau_{p0} L b \tag{7-13}$$

and that for the F-F from

$$F_0 = \tau_{p0} b\left(\frac{2L_0^2 E}{\tau_{p0} b}\right)^{1/3}, \tag{7-14}$$

where L is the average spacing of impurities along the dislocation, L_0 the average spacing of impurities on the slip plane, and E the line tension of the dislocations. τ_{p0} is given in table 7-3. The value of T_c in equation (7-11) is obtained from figure 7-6 (a) and that in equation (7-12) from figure 7-6 (b). The relation between (τ_{p1}/τ_{p0}) and temperature is considered to reveal the force-distance relation between a dislocation and an impurity, because τ_{p1} corresponds to the effective stress due to only one type of the impurities which lie on the dislocation when the dislocation moves forward with the help of oscillation (see sec. 2-4 to 2-7 of chap. 2).

Table 7-3. Values of τ_{p0} for various force-distance relations between a dislocation and the impurity in the two specimens

Single crystal	force-distance relation	τ_{p0} (MPa)
Quenched specimen	Fleischer	14.52
	F-F	25.47
Annealed specimen	SQ	2.17
	PA	3.28
	TR	4.53

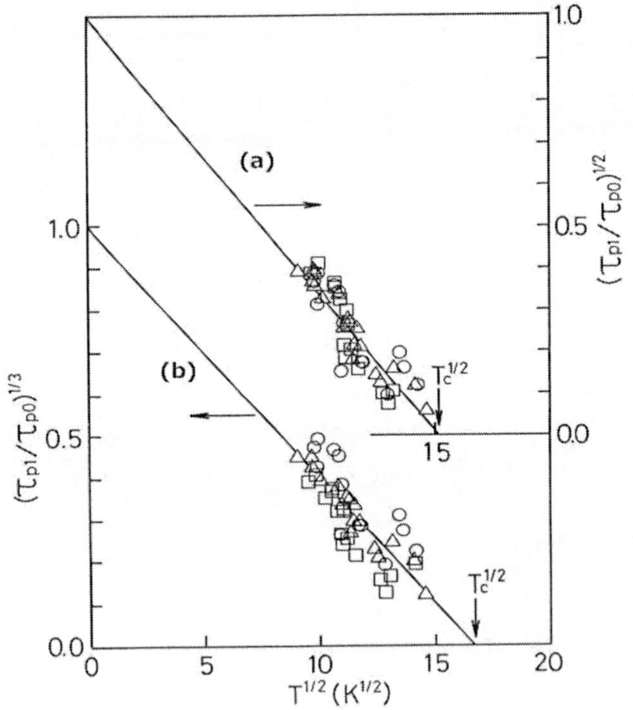

Figure 7-6. Linear plots of the effective shear stress and the temperature at (a) the Fleischer's model and (b) the F-F for the quenched specimen: KCl:Sr^{2+} ((○) 0.035 mol.%, (△) 0.050 mol.%, (□) 0.065 mol.% in the melt).

The open circles are fitted to the solid line in comparison with the dashed line as shown in figure 7-5. Therefore, the force-distance relation for the quenched specimen is suited to the F-F rather than the Fleischer's model at the temperature. Even though the open circles are given within the narrow range of temperature, this method is useful for the determination of suitable model.

7.2.3.2. Annealed Specimen

It was clear that the force-distance relation for the annealed specimen could not be approximated by the Fleischer's model and the F-F (see sec. 4.5 of chap. 4 and sec. 7.2.2 of chap. 7), whereas the other models were not distinct in section 7.2.2 of this chapter. One investigates the applicability of three models [4]: a square, a parabolic, and a triangular force-distance relation, which are termed the SQ, the PA, and the TR respectively in this chapter. The three force-distance relations are taken into account the Friedel relation [3].

When the thermally activated overcoming of the aggregates controls the dislocation velocity and the dislocation moves forward the distance of L_0^2/L, the dislocation velocity is given by

$$\upsilon = \nu\left(\frac{L_0^2}{L}\right)\exp\left(\frac{-\Delta G}{kT}\right), \tag{7-15}$$

where ν is the frequency of vibration of a dislocation segment, ΔG is the change in Gibbs free energy of activation for overcoming of local barriers by dislocations. The Gibbs free energy for the SQ is given by (combination of equations (5-12) and (7-46))

$$\Delta G = \Delta G_0 - \beta\tau^{*2/3} \tag{7-16}$$

$$\beta = \left(2\mu b^4 d^3 L_0^2\right)^{1/3}, \tag{7-17}$$

where μ is the shear modulus and d is an activation distance. The PA gives (see the subsequent sec. 7.3 of chap. 7)

$$\Delta G = \Delta G_0\left\{1-\left(\frac{\tau^*}{\tau_0^*}\right)^{2/3}\right\}^{3/2} \tag{7-18}$$

and the TR gives (see the subsequent sec. 7.3 of chap. 7)

$$\Delta G = \left(\frac{\beta}{8}\right)\left(\tau_0^{*2/3} - 2\tau^{*2/3} + \tau_0^{*-2/3}\tau^{*4/3}\right). \tag{7-19}$$

Combining equation (7-17) and the Friedel relation:

$$L = \left(\frac{2L_0^2 E}{\tau_0^* b}\right)^{1/3}, \tag{7-20}$$

the β is expressed by

$$\beta = \tau_0^{*1/3} Lbd = \Delta G_0 \tau_0^{*-2/3}. \tag{7-21}$$

Substituting equation (7-21) into equations (7-16) and (7-19), we find

$$\Delta G = \Delta G_0 \left\{ 1 - \left(\frac{\tau^*}{\tau_0^*} \right)^{2/3} \right\} \tag{7-22}$$

and

$$\Delta G = \left(\frac{\Delta G_0}{8} \right) \left\{ 1 - 2\left(\frac{\tau^*}{\tau_0^*} \right)^{2/3} + \left(\frac{\tau^*}{\tau_0^*} \right)^{4/3} \right\}. \tag{7-23}$$

From equations (7-15) and (7-22), the dislocation velocity for the SQ is expressed by

$$\upsilon = \nu \left(\frac{L_0^2}{L} \right) \exp\left[\frac{-F_0 b \left\{ 1 - \left(\tau^* / \tau_0^* \right)^{2/3} \right\}}{kT} \right]. \tag{7-24}$$

From equations (7-15) and (7-18), that for the PA yields

$$\upsilon = \nu \left(\frac{L_0^2}{L} \right) \exp\left[\frac{-F_0 b \left\{ 1 - \left(\tau^* / \tau_0^* \right)^{2/3} \right\}^{3/2}}{kT} \right]. \tag{7-25}$$

Similarly from equations (7-15) and (7-23), that for the TR is

$$\upsilon = \nu \left(\frac{L_0^2}{L} \right) \exp\left[\frac{-F_0 b \left\{ 1 - 2\left(\tau^* / \tau_0^* \right)^{2/3} + \left(\tau^* / \tau_0^* \right)^{4/3} \right\}}{8kT} \right]. \tag{7-26}$$

It is well known that m^* can be obtained from the following equation:

$$m^* = \frac{\partial \ln \upsilon}{\partial \ln \tau^*} = \tau^* \left(\frac{\partial \ln \upsilon}{\partial \tau^*} \right). \tag{7-27}$$

$(\partial \ln \upsilon / \partial \tau^*)$ in equation (7-27) can be found from equation (7-25) as follows

$$\frac{\partial \ln \upsilon}{\partial \tau^*} = \left(\frac{2F_0 b}{3kT\tau_0^*} \right) \left(\frac{\tau^*}{\tau_0^*} \right)^{-1/3}. \tag{7-28}$$

From equation (7-25), we can find

$$\frac{\partial \ln \upsilon}{\partial \tau^*} = \left(\frac{F_0 b}{kT\tau_0^*} \right) \left\{ 1 - \left(\frac{\tau^*}{\tau_0^*} \right)^{2/3} \right\}^{1/2} \left(\frac{\tau^*}{\tau_0^*} \right)^{-1/3}. \tag{7-29}$$

Similarly from equation (7-26), we obtain

$$\frac{\partial \ln \upsilon}{\partial \tau^*} = \left(\frac{F_0 b}{6kT\tau_0^*}\right)\left\{\left(\frac{\tau^*}{\tau_0^*}\right)^{-1/3} - \left(\frac{\tau^*}{\tau_0^*}\right)^{1/3}\right\}. \tag{7-30}$$

Therefore, m^* for the SQ is expressed by

$$m^* = \left(\frac{2F_0 b}{3kT}\right)\left(\frac{\tau^*}{\tau_0^*}\right)^{2/3}. \tag{7-31}$$

The formula relating the effective stress and temperature for the SQ is

$$\left(\frac{\tau^*}{\tau_0^*}\right)^{2/3} = 1 - \left(\frac{T}{T_c}\right). \tag{7-32}$$

For PA the equivalent expression is

$$\left(\frac{\tau^*}{\tau_0^*}\right)^{2/3} = 1 - \left(\frac{T}{T_c}\right)^{2/3} \tag{7-33}$$

and for TR

$$\left(\frac{\tau^*}{\tau_0^*}\right)^{2/3} = 1 - \left(\frac{T}{T_c}\right)^{1/2}. \tag{7-34}$$

Substituting equation (7-32) into equation (7-31), m^* for the SQ can be obtained as follows

$$m^* = \left(\frac{2F_0 b}{3k}\right)\left(T^{-1} - T_c^{-1}\right). \tag{7-35}$$

In the same way, m^* for the PA from equations (7-27), (7-29), and (7-33) is expressed by

$$m^* = \left(\frac{F_0 b}{kT}\right)\left\{\left(\frac{T}{T_c}\right)^{1/3} - \left(\frac{T}{T_c}\right)\right\} \tag{7-36}$$

and using equations (7-27), (7-30), and (7-34), m^* for the TR is

$$m^* = \left(\frac{F_0 b}{6kT}\right)\left\{\left(\frac{T}{T_c}\right)^{1/2} - \left(\frac{T}{T_c}\right)\right\}. \tag{7-37}$$

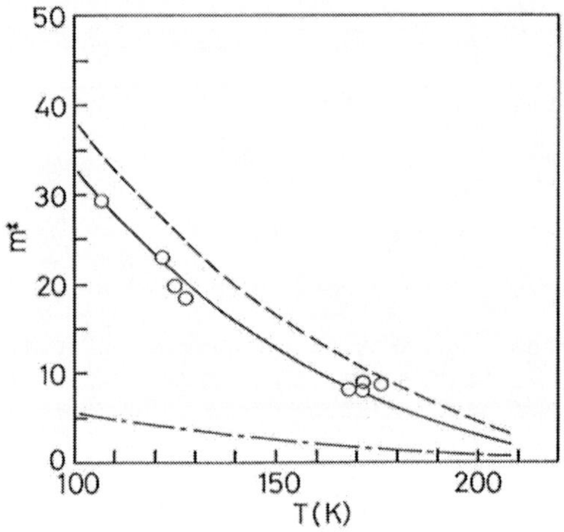

Figure 7-7. Relationship between the temperature and the dislocation velocity-effective stress exponent for the annealed specimen at the three models: (- - -) the PA, (—) the SQ, and (–·–) the TR. Open circles represent the $m_{\varepsilon=0}$ for the annealed specimen.

Each curve in figure 7-7 for the annealed specimen corresponds to the relation between temperature and m^* derived from equation (7-35) for the SQ, equation (7-36) for the PA, and equation (7-37) for the TR. F_0 for the three models is calculated from equation (7-14) in which the value of τ_{p0} is given in table 7-3. F_0 and T_c are tabulated in table 7-2. The relative curve of temperature and the m^* for the SQ, that for the PA, and that for the TR are represented as a solid line, a dashed line, and a dash-dotted line in figure 7-7. Open circles correspond to the m^* for the annealed specimen,which are obtained from $m_{\varepsilon=0}$ .Unfortunately, it was difficult to evaluatem^* from the measurement of $m_{\varepsilon=0}$ above around 180 K. As can be seen from figure 7-7, the open circles increase with decreasing temperature and approach to the solid line of SQ. Accordingly, it is considered that the SQ can be selected as the most suitable force-distance relation among the three.

7.2.4. F_0, L/L_0, and Φ_0

By using equation (7-14), F_0 was provided in the preceding section 7.2.3.1. Then, the line tension of the dislocations is calculated by μb^2. The shear modulus is assumed to be 1.01×10^{10} Pa for [110] direction at 0 K [16]. Further, the average spacing of isolated I-V dipoles on the slip plane may be given by [2, 17]

$$L_0 = \frac{b}{\left(\frac{4c_1}{3}\right)^{1/2}}, \tag{7-38}$$

where the concentration of the impurities (c_1) is 98.3 ppm from dielectric loss measurement. The average spacing of the aggregates on the slip plane is given by

$$L_0 = \frac{b}{c_2^{1/2}}, \tag{7-39}$$

where the concentration of the aggregates (c_2) is 32.1 ppm from atomic absorption method, because the aggregates in the annealed specimen are assumed to form the trimers which are arranged hexagonally head to tail in a (111) planes by the heat treatment [18]. As a result, F_0 calculated from equation (7-14) is 9.20×10^{-10} N for the quenched specimen and is 2.85×10^{-10} N for the annealed specimen. Thus F_0 is reduced to about one-third by the heat treatment. In addition, the average spacing (L) of impurities along the dislocation at 0 K is 812 Å for the quenched specimen and 2947 Å for the annealed specimen on a calculation from the values of F_0 and τ_{p0}. The L is about twice as long as L_0 for the quenched specimen. On the other hand it is the quadruple of L_0 for the annealed specimen. Therefore, the bending angle (ϕ_0) at which the dislocation embraces the aggregates under the τ_{p0} is expected to be larger than that for the quenched specimen. We further examine the ϕ_0 for both the specimens. From the equation (see sec. 5.4 of chap. 5)

$$\frac{\phi_0}{2} = \cos^{-1}\left(\frac{\tau_{p0} L_0 b}{2E}\right)^{2/3}. \tag{7-40}$$

The ϕ_0 for the quenched specimen is 154 degrees and that for the annealed specimen 172 degrees. The various characteristics are summarized in table 7-4. L_0 becomes larger after the heat treatment. That is, the concentration of weak obstacles such as impurities decreases after the treatment. In this section, it is found that F_0 decreases and ϕ_0 increases on annealing of the quenched specimen. Consequently, the resistance of obstacle to the dislocation movement in the quenched specimen is weakened by annealing it. The interaction between a dislocation and the impurity for the quenched specimen is stronger in comparison with that for the annealed specimen.

Table 7-4. Values of various characteristics[a] for the two specimens

Single crystal	F_0 ($\times10^{-10}$ N)	L_0 (Å)	L/L_0	ϕ_0 (degrees)
Quenched specimen	9.20	389	2.1	154
Annealed specimen	2.85	786	3.8	172

[a] The interaction between a dislocation and the impurity in the quenched specimen is approximated by the F-F and that in the annealed specimen by the SQ.

7.2.5. Influence of Heat Treatment on *m**

7.2.5.1. m Obtained on the Basis of* $m_{\varepsilon=0}$

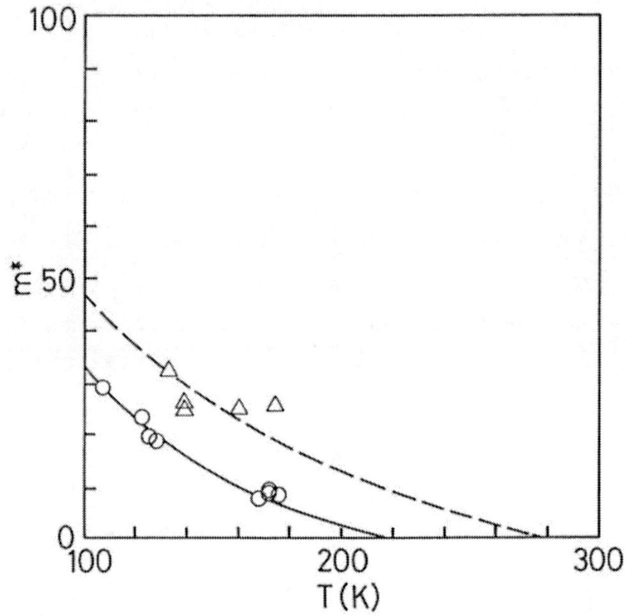

Figure 7-8. Temperature dependence of the dislocation velocity-effective stress exponent for the two specimens: (△) the quenched specimen and (○) the annealed specimen. Open symbols represent $m_{\varepsilon=0}$. The dashed and the solid curves are calculated by equations (7-12) and (7-35), respectively.

As described in the section 7.2.4 of this chapter 7, F_0 decreases and ϕ_0 increases on annealing of the quenched specimen. This means that the resistance of obstacle to the dislocation movement in the quenched specimen is weakened by annealing it. In addition, L_0 becomes large after the heat treatment. That is, the concentration of weak obstacles such as impurities decreases after the treatment. If these results for F_0, L/L_0, ϕ_0, and L_0 in section 7.2.4 are real, it is expected that the value of m^* for KCl:Sr^{2+} becomes smaller by the heat treatment according as the description [19] : m^* becomes smaller when the concentration or the resistance of obstacle to dislocation motion decreases. Figure 7-8 shows the m^* for the two kinds of specimens. Open triangles and circles correspond to $m_{\varepsilon=0}$ for the quenched specimen and for the annealed specimen, obtained in the section 7.2.3. Dashed and solid lines are derived from equations (7-12) and (7-35). Equation (7-12) represents m^* for the Fleischer's model taking account of the Friedel relation and equation (7-35) does that for a square force-distance relation taking account of the Friedel relation. It is clear that the m^* for the annealed specimen is less than that for the quenched specimen at a given temperature, thus proving the above-mentioned prediction, i.e., m^* decreases by the heat treatment. The difference between the values of m^* for the two specimens is 13 to 17 within the temperatures. Figure 7-8 also suggests that the dislocation velocity in the quenched specimen is more sensitive to the effective stress due to the obstacles than that in the annealed specimen.

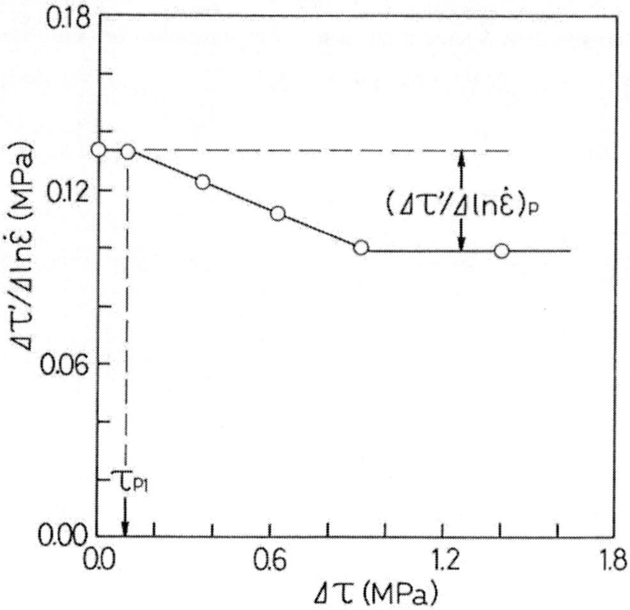

Figure 7-9. Relationship between the strain-rate sensitivity and the stress decrement for the annealed specimen at 201 K and ε=18%.

7.2.5.2. m Obtained on the Basis of the Relation between Strain-Rate Sensitivity and $\Delta\tau$*

We deduce the values of m^* from the relative curve of strain-rate sensitivity and stress decrement due to the oscillation. The curve has two bending points and two plateau regions and strain-rate sensitivity decreases with $\Delta\tau$ between the two bending points. Figure 7-9 shows the typical variation of strain-rate sensitivity with $\Delta\tau$. The curve is considered to represent the influence of ultrasonic oscillation on the dislocation motion on the slip plane containing many weak obstacles such as impurities and a few strong ones such as forest dislocations (see sec. 2.7 of chap. 2). $(\Delta\tau'/\Delta\ln\dot{\varepsilon})_p$, which is given by the difference between strain-rate sensitivities at first plateau place and at second one on the relative curve of strain-rate sensitivity and $\Delta\tau$, is assumed to be the strain-rate sensitivity due to impurities when the dislocation moves forward with the help of oscillation (see sec. 2.1 of chap. 2 and sec. 4.4 of chap. 4). $\Delta\tau'$ is the stress change due to the strain-rate cycling when the strain-rate cycling associated with the oscillation is carried out keeping the stress amplitude constant. Then, m^* may be expressed by

$$m^* = \tau_{p1}\left(\frac{\Delta\ln\dot{\varepsilon}}{\Delta\tau'}\right)_p, \qquad (7\text{-}41)$$

where m^* is examined on the assumption that equation (7-27) is also valid for the case of applying the ultrasonic oscillatory stress during plastic deformation. The results of equation

(7-41) are indicated by open symbols in figure 7-10. Two curves in the figure are to guide the reader's eye. The values of $\tau_{p1}(\Delta\ln\dot{\varepsilon}/\Delta\tau')_p$ tend to increase with decreasing temperature as well as $m_{\varepsilon=0}$ shown in figure 7-8. Furthermore, $\tau_{p1}(\Delta\ln\dot{\varepsilon}/\Delta\tau')_p$ for the annealed specimen is less than that for the quenched specimen at a given temperature. This is also similar to the above observations of $m_{\varepsilon=0}$ for the two specimens.

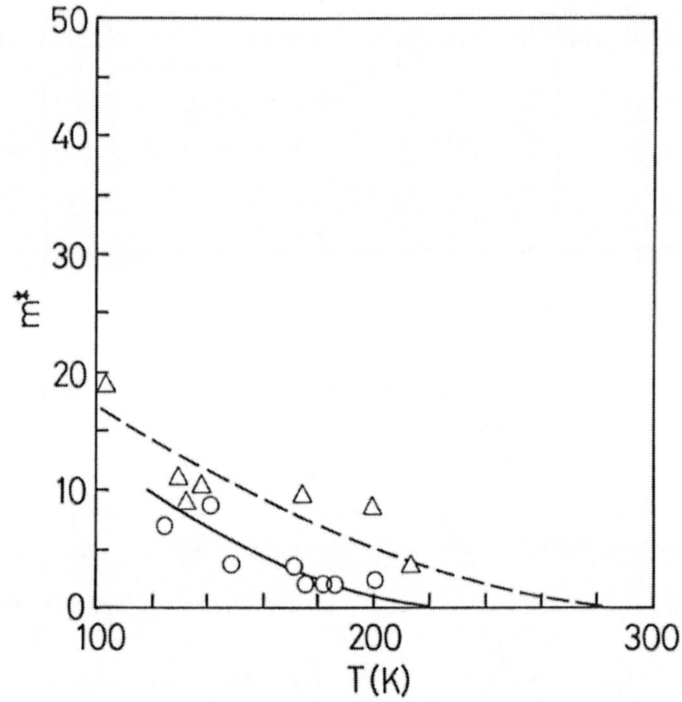

Figure 7-10. Temperature dependence of the dislocation velocity-effective stress exponent for the two specimens: (Δ) the quenched specimen and (○) the annealed specimen. Open symbols represent m^* evaluated by $\tau_{p1}(\Delta\ln\dot{\varepsilon}/\Delta\tau')_p$.

7.2.6. Proportionality of Δ*H*(*T*)

The relation between the temperature and the activation enthalpy is investigated for the interaction between a dislocation and the aggregate in the annealed specimen. The activation enthalpy for the SQ is calculated from the following relation:

$$\Delta H = kT^2\left(\frac{\Delta\ln\dot{\varepsilon}}{\Delta\tau'}\right)_p \frac{3\tau_{p0}\{1-(T/T_c)\}^{1/2}}{2T_c}. \tag{7-42}$$

The Δ*H* for the PA is expressed by

$$\Delta H = kT^2 \left(\frac{\Delta \ln \dot{\varepsilon}}{\Delta \tau'} \right)_p \frac{\tau_{p0} \left\{ \left(T_c / T \right)^{2/3} - 1 \right\}^{1/2}}{T_c} \tag{7-43}$$

and also, that for the TR is expressed by

$$\Delta H = kT^2 \left(\frac{\Delta \ln \dot{\varepsilon}}{\Delta \tau'} \right)_p \frac{3\tau_{p0} \left\{ \left(T_c / T \right) - \left(T_c / T \right)^{1/2} \right\}^{1/2}}{4T_c} . \tag{7-44}$$

The processes of leading equations (7-42) to (7-44) are described in the subsequent sections 8-6 and 8-7 of chapter 8. The results of calculations for $\Delta H(T)$ are shown in figure 7-11 (a) for the SQ, figure 7-11 (b) for the PA, and figure 7-11 (c) for the TR, respectively. The lines of $\Delta H(T)$ are determined by the method of least squares. However, it is difficult to select the most suitable force-distance relation among the three from the proportionality of $\Delta H(T)$ within the temperatures as shown in figure 7-11 (a) to (c). The value of $\Delta H(T_c)$ obtained from these figures is given in table 7-5. The value of $\Delta H(T_c)$ is almost the same for the three force-distance relations from table 7-5.

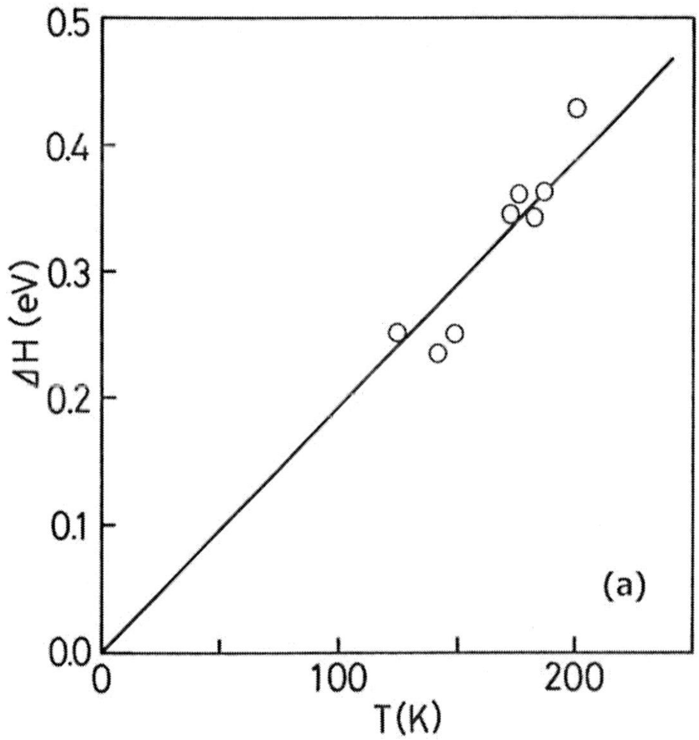

Figure 7-11. (Continued).

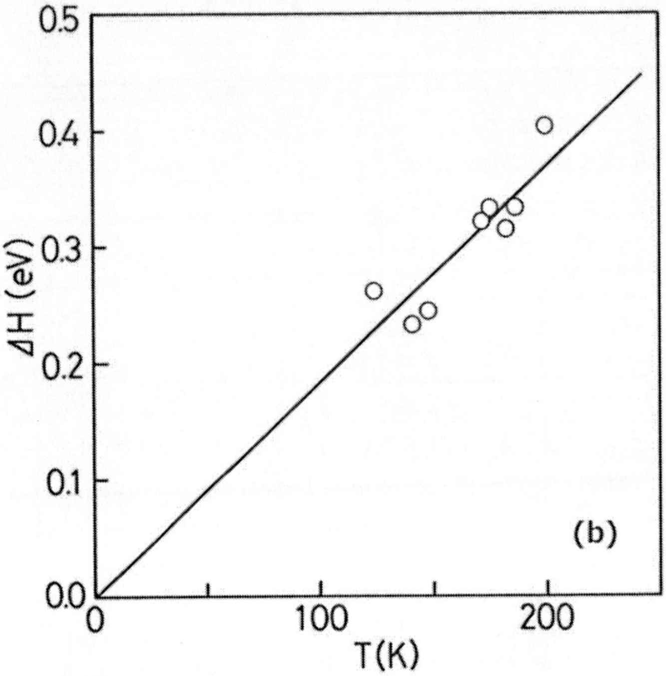

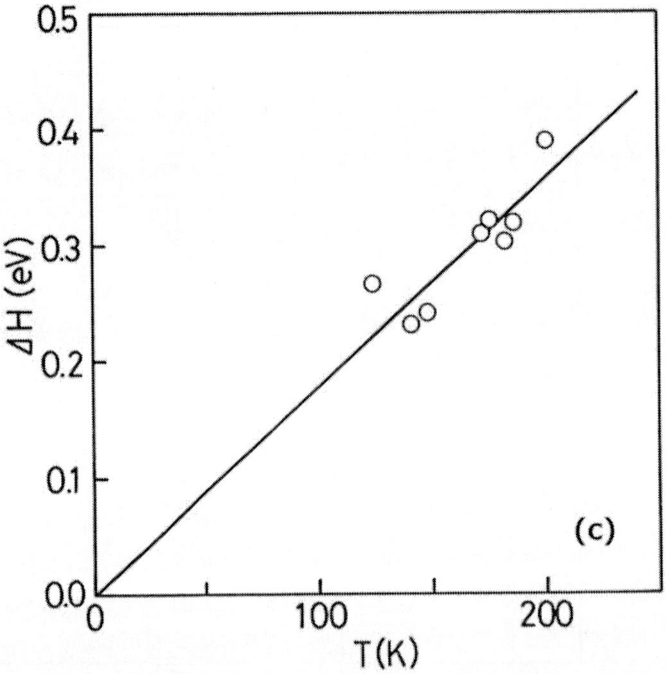

Figure 7-11. Proportional relationship between the temperature and the activation enthalpy for the interaction between a dislocation and the aggregate in the annealed specimen at various models: (a) the SQ, (b) the PA, and (c) the TR.

Table 7-5. Values of $\Delta H(T_c)$ for various force-distance relations between a dislocation and the aggregate in the annealed specimen

The force-distance relation	$\Delta H(T_c)$ (eV)
SQ	0.43
PA	0.41
TR	0.41

7.3. The Gibbs Free Energy for Overcoming of the Aggregate by a Dislocation

The thermally activated deformation rate ($\dot{\varepsilon}$) is expressed by an Arrhenius-type equation:

$$\dot{\varepsilon} = \dot{\varepsilon}_0 \exp\left(\frac{-\Delta G}{kT}\right), \tag{7-45}$$

where $\dot{\varepsilon}_0$ is a frequency factor and is unique for a particular dislocation mechanism. The change in Gibbs free energy is expressed for square force-distance relation between a dislocation and an impurity as follows

$$\Delta G = \Delta G_0 - \tau^* Lbd\,, \tag{7-46}$$

where L is the length of dislocation and d is an activation distance. Differentiating the substitutional equation of equations (7-16) and (7-17) in equation (7-45) with respect to the shear stress, we find

$$\tau_{p1}\left(\frac{\partial \ln \dot{\varepsilon}}{\partial \tau^*}\right) = \left(\frac{2\Delta G_0}{3kT}\right) + \frac{2}{3}\ln\left(\frac{\dot{\varepsilon}}{\dot{\varepsilon}_0}\right). \tag{7-47}$$

Two other models, namely the PA and the TR, will be examined. The force (F) acted on the dislocation as a function of the distance (x) until the obstacle on the slip plane is

$$F(x) = F_0\left(1 - \frac{|x|^n}{a^n}\right),\ |x| < a \qquad (F(\pm a) = 0,\ F(0) = F_0) \tag{7-48}$$

where triangular $F(x)$ is for the case of n=1 and parabolic $F(x)$ for n=2 [16]. The representation of figure 7-12 is concerned with a common type of thermal activation barrier and corresponds to typical force-distance curve for short-range obstacle. Shape of this force-distance curve represented by $F(x)$ means the model overcoming the obstacle by a dislocation.

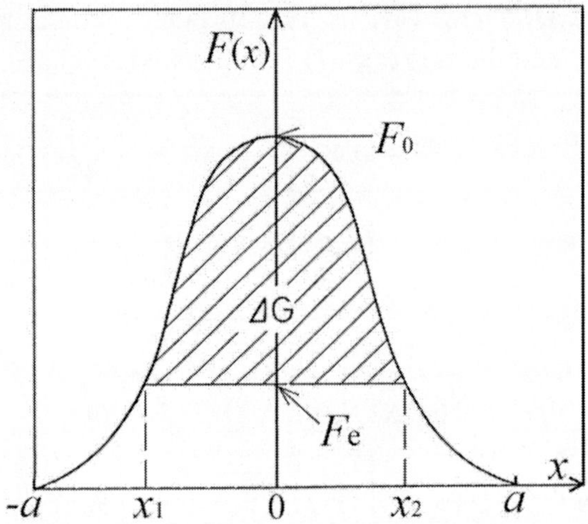

Figure 7-12. The process for thermal activated overcoming of the short-range obstacle by a dislocation. Variation in the force, $F(x)$, acted on the dislocation with the distance, x, for a dislocation motion.

When a dislocation overcomes the short-range obstacles by the force F_e, ΔG, which is taken as the shaded area under $F(x)$, is the Gibbs free energy of activation for the breakaway of the dislocation from the obstacle. Then, ΔG is calculated by

$$\Delta G = \int_{-a}^{a} F(x)dx - \int_{-a}^{x_1} F(x)dx - \int_{x_2}^{a} F(x)dx - \int_{x_1}^{x_2} F_e dx . \tag{7-49}$$

Equation (7-49) is transformed to the following equation:

$$\Delta G = 2\left\{ \int_{0}^{1} F_0 a(1-X^n)dX - \int_{X}^{1} F_0 a(1-X^n)dX - \int_{0}^{X} F_e a dX \right\}, \ (X = x/a). \tag{7-50}$$

Using equation (7-48), X in equation (7-50) is given by

$$X = \left(1 - \frac{F_e}{F_0}\right)^{1/n} . \tag{7-51}$$

Substituting of equation (7-51) in equation (7-50), the Gibbs free energy is given by

$$\Delta G = 2F_0 a\left(\frac{n}{n+1}\right)\left(1 - \frac{F}{F_0}\right)^{(n+1)/n} , \tag{7-52}$$

where F_e is replaced by F.

The Gibbs free energy for the PA obtained from the Friedel relation (i.e., equation (7-20) in which τ^*_0 is replaced by τ_{p0}) and equation (7-52) is expressed by

$$\Delta G = \Delta G_0 \left\{ 1 - \left(\frac{\tau_{p1}}{\tau_{p0}} \right)^{2/3} \right\}^{3/2}, \tag{7-53}$$

where τ^* and τ^*_0 are replaced by τ_{p1} and τ_{p0}, respectively. Differentiating the substitutional equation of equation (7-50) in equation (7-45) with respect to the shear stress, we find

$$\frac{\partial \ln \dot{\varepsilon}}{\partial \tau^*} = \frac{\Delta G_0}{kT} \left\{ \left(\tau_{p0}^2 \tau_{p1} \right)^{-2/3} - \tau_{p0}^{-2} \right\}^{1/2} + \frac{\partial \ln \dot{\varepsilon}_0}{\partial \tau^*}. \tag{7-54}$$

From equations (7-20) and (7-52), the Gibbs free energy for the TR yields

$$\Delta G = \frac{\beta}{8} \left(\tau_{p0}^{2/3} - 2\tau_{p1}^{2/3} + \tau_{p0}^{-2/3} \tau_{p1}^{4/3} \right), \ \left(\beta = (2\mu b^4 a^3 L_0^2)^{1/3} \right) \tag{7-55}$$

where τ^* and τ^*_0 are replaced by τ_{p1} and τ_{p0}. Similarly from equations (7-45) and (7-55), we find

$$\frac{\partial \ln \dot{\varepsilon}}{\partial \tau^*} = \frac{4\Delta G_0}{3kT} \left\{ \left(\tau_{p0}^2 \tau_{p1} \right)^{-1/3} - \left(\tau_{p0}^{-4} \tau_{p1} \right)^{1/3} \right\} + \frac{\partial \ln \dot{\varepsilon}_0}{\partial \tau^*}. \tag{7-56}$$

The results of calculations for equations (7-47), (7-54) and (7-56) are shown as the open circles in figure 7-13 (a), (b) and (c), respectively. The Gibbs free energy for the interaction between a dislocation and the aggregate in the annealed specimen, which is obtained from the slopes of straight lines in figure 7-13 (a) to (c), is given in table 7-6.

Table 7-6. Values of ΔG_0 for various force-distance relations between a dislocation and the aggregate in the annealed specimen

The force-distance relation	ΔG_0 (eV)
SQ	0.26
PA	0.18
TR	0.17

The SQ seems to be the most suitable force-distance relation among the three from figures 7-3 and 7-7. Then, the width of force-distance curve for the SQ without applied stress can be obtained from the following equation:

$$d = \frac{\Delta G_0}{\tau_{p0} b} \left(\frac{\tau_{p0} b}{2L_0^2 E} \right)^{1/3}, \tag{7-57}$$

where the line tension of the dislocations is calculated by μb^2. The shear modulus for [110] direction is assumed to be 1.01×10^{10} Pa at 0 K [16]. The average spacing of the aggregates on the slip plane is given by

$$L_0 = b/c^{1/2}, \tag{7-58}$$

where the concentration of the aggregates, c, is 32.1 ppm from atomic absorption method. As a result, the width of the force-distance curve was found to be 1.5 Å less than the magnitude of the Burgers vector.

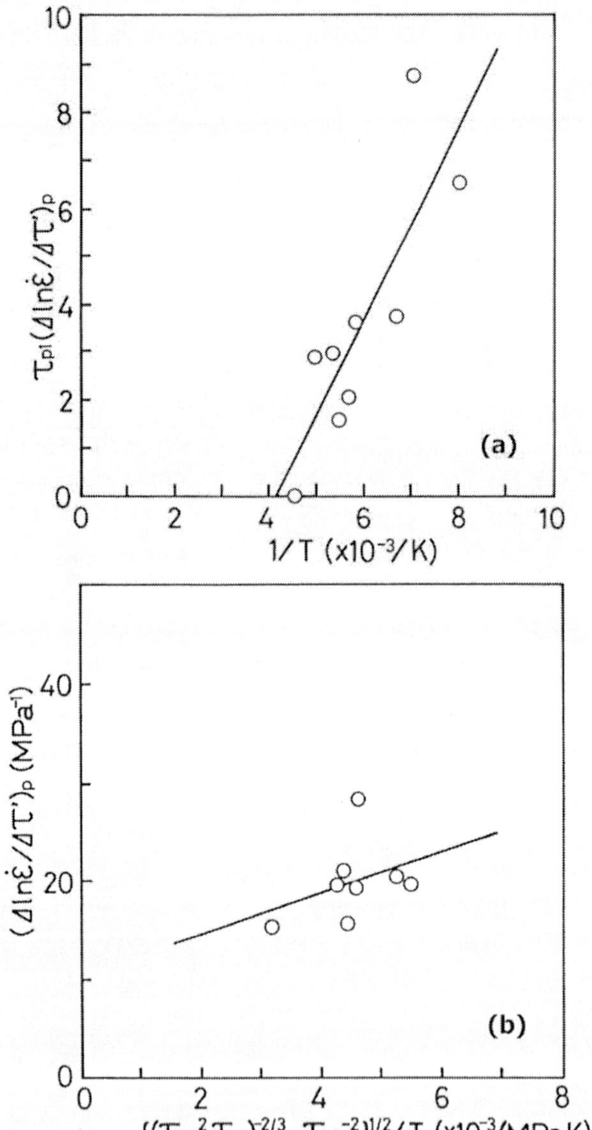

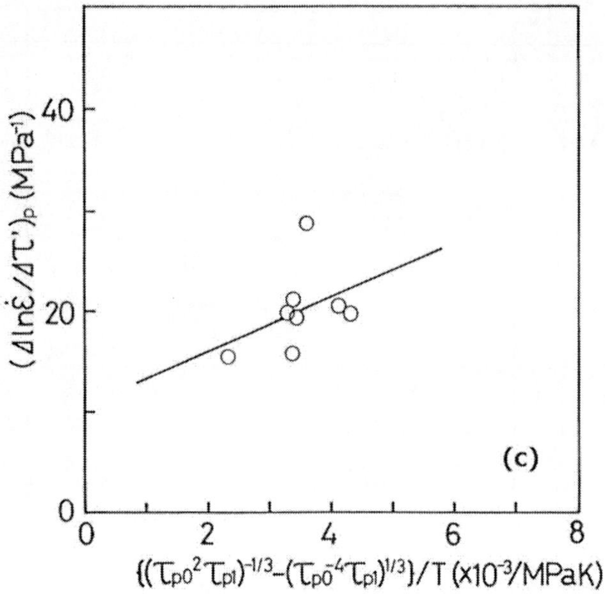

Figure 7-13. Linear plots of (a) equation (7-47) for the SQ, (b) equation (7-54) for the PA, and (c) equation (7-56) for the TR in the specimen.

7.4. Influence of the State of Impurities Due to the Heat Treatment on the Activation Energy for Breakaway from the Impurity by a Dislocation

7.4.1. Activation Enthalpy for the Dislocation Motion

The ΔH for the breakaway of a dislocation from the isolated I-V dipole in the quenched specimen can be calculated for the F-F by using the equation (see sec. 5.6 of chap. 5):

$$\Delta H = -kT^2\left(\frac{\Delta \ln \dot{\varepsilon}}{\Delta \tau'}\right)_p \left(\frac{-3\tau_{p0}}{2T_c}\right)\left(\frac{T_c}{T}\right)^{1/2}\left\{1-\left(\frac{T}{T_c}\right)^{1/2}\right\}^2 . \tag{7-59}$$

As for the annealed specimen, the ΔH is obtained from equation (7-42) for the SQ. The values of ΔH derived from equation (7-59) are plotted as a function of temperature in figure 7-14 for the quenched specimen, while those from equation (7-42) are plotted in figure 7-15 for the annealed specimen. The values of $\Delta H(T_c)$ taken from figures 7-14 and 7-15, which correspond to the activation enthalpy for breaking away from the strain field around the impurity by a dislocation at 0 K, are given in table 7-7. The activation enthalpy becomes small for the annealed specimen, compared with the quenched one.

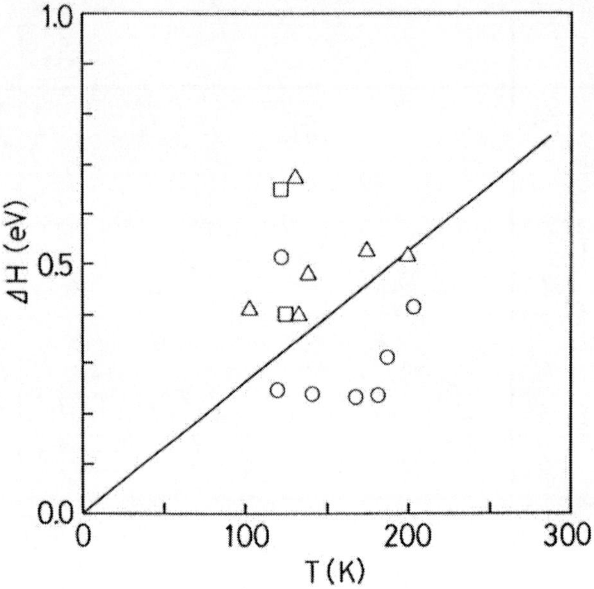

Figure 7-14. Proportional relationship between the temperature and the activation enthalpy for the interaction between a dislocation and the impurity in the quenched specimen: KCl:Sr^{2+} ((○) 0.035 mol.%, (△) 0.050 mol.%, (□) 0.065 mol.% in the melt).

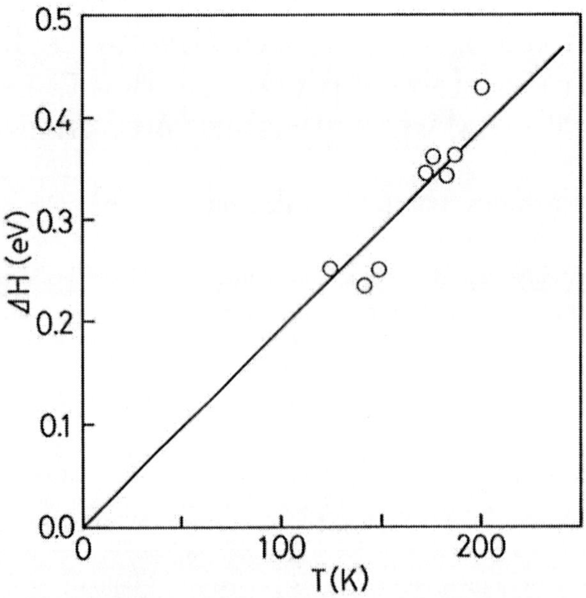

Figure 7-15. Proportional relationship between the temperature and the activation enthalpy for the interaction between a dislocation and the aggregate in the annealed specimen.

Table 7-7. Values of $\Delta H(T_c)$ for the quenched and the annealed specimens

Single crystal	$\Delta H(T_c)$ (eV)
Quenched specimen	0.79
Annealed specimen	0.43

7.4.2. Gibbs Free Energy for the Dislocation Motion

The Gibbs free energy for the interaction between a dislocation and the impurity is expressed from an Arrhenius-type equation for the thermally activated deformation rate at low temperature as the equation (7-45), namely

$$\Delta G = \alpha kT , \left(\alpha = \ln\left(\frac{\dot{\varepsilon}_0}{\dot{\varepsilon}} \right) \right) \tag{7-60}$$

and the ΔG for the F-F is also expressed by (see sec. 5.3 of chap. 5)

$$\Delta G = \Delta G_0 \left\{ 1 - \left(\frac{\tau^*}{\tau_0^*} \right)^{1/3} \right\}^2 , \ (\Delta G_0 = F_0 b). \tag{7-61}$$

Differentiating the combination of equations (7-60) and (7-61) with respect to the shear stress gives

$$\frac{\partial \ln \dot{\varepsilon}}{\partial \tau^*} = \left(\frac{2\Delta G_0}{3kT\tau_0^*} \right) \left(\frac{\tau_0^*}{\tau^*} \right)^{2/3} \left\{ 1 - \left(\frac{\tau^*}{\tau_0^*} \right)^{1/3} \right\} + \frac{\partial \ln \dot{\varepsilon}_0}{\partial \tau^*} . \tag{7-62}$$

Further, substituting equation (7-1) in equation (7-62) gives

$$\frac{\partial \ln \dot{\varepsilon}}{\partial \tau^*} = \left(\frac{2\Delta G_0}{3kT\tau_{p0}} \right) \left\{ 1 - \left(\frac{T}{T_c} \right)^{1/2} \right\}^{-2} \left(\frac{T}{T_c} \right)^{1/2} + \frac{\partial \ln \dot{\varepsilon}_0}{\partial \tau^*} , \tag{7-63}$$

where τ^*_0 is replaced by τ_{p0}. The result of calculations of equation (7-63) for the F-F is shown in figure 7-16 for the quenched specimen. The open symbols correspond to the $(\Delta \ln \dot{\varepsilon} / \Delta \tau')_p$ for the quenched specimen. On the basis of the slope of straight line, the Gibbs free energy is 0.39 eV. The Gibbs free energy for the interaction between a dislocation and the aggregate in the annealed specimen, which is obtained through the slope of straight line in figure 7-17, is 0.26 eV. The open circles in figure 7-17 show the results of calculations for equation (7-47). The Gibbs free energy for the annealed specimen is smaller than that of the quenched one as tabulated in table 7-8.

Table 7-8. Values of ΔG_0 for the quenched and the annealed specimens

Single crystal	ΔG_0 (eV)
Quenched specimen	0.39
Annealed specimen	0.26

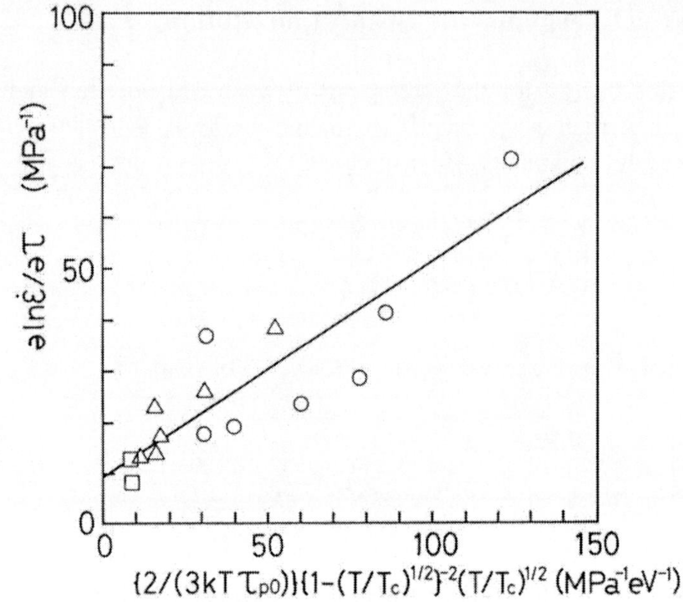

Figure 7-16. Linear plots of equation (7-63) for the quenched specimen: KCl:Sr^{2+} ((○) 0.035 mol.%, (Δ) 0.050 mol.%, (□) 0.065 mol.% in the melt).

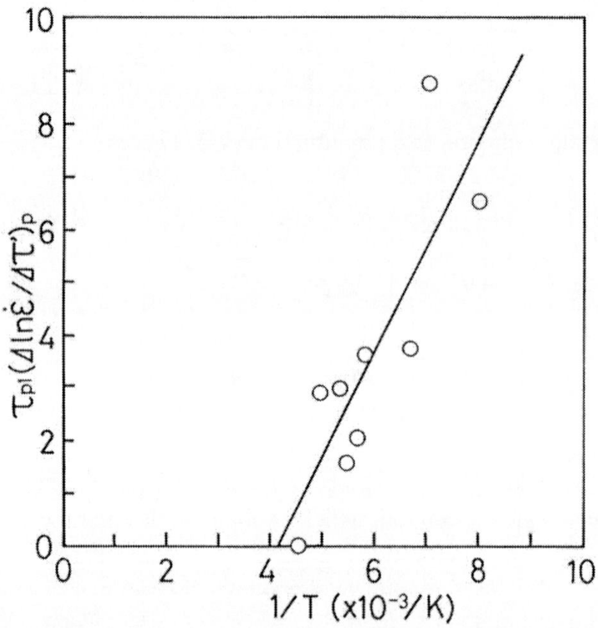

Figure 7-17. Linear plots of equation (7-47) for the annealed specimen. (This figure is the same as figure 7-13 (a).)

CONCLUSION

The following conclusions 1 to 5 were derived in this chapter by the strain-rate cycling tests associated with the ultrasonic oscillation.

1. The most suitable model between a dislocation and the aggregate in the annealed specimen cannot be selected among the three (the SQ, the PA, and the TR) from the linear effective stress-temperature relation or from the proportionality of $\Delta H(T)$ within the temperature. Then, the $\Delta H(T_c)$ is almost the same for the three force-distance relations.
2. On the basis of the dependence of the strain-rate sensitivity due to the aggregates on temperature, it was found that the interaction between a dislocation and the aggregate in the annealed specimen cannot be approximated to the F-F. The SQ seemed to be the most suitable model among the three which are the SQ, the PA, and the TR.
3. On the basis of m^* obtained by $m_{\varepsilon=0}$, the force-distance relation between a dislocation and the impurity in the quenched specimen is approximated by the F-F rather than the Fleischer's model. As for the annealed specimen, the SQ is the most appropriate among the SQ, the PA and the TR models. In spite of the narrow range of temperature, this method is useful for the determination of suitable force-distance relation.
4. The resistance of weak obstacle to the dislocation movement in the quenched specimen is weakened by annealing it (e.g., F_0 is reduced to about one-third and ϕ_0 increases from 154 to 172 degrees). Furthermore, the concentration of weak obstacles decreases after the heat treatment. These may lead to the results that m^* calculated by $m_{\varepsilon=0}$ or by $\tau_{p1}(\Delta \ln \dot{\varepsilon} / \Delta \tau')_p$ at a given temperature is lowered by the heat treatment. That is to say, the dislocation velocity in the quenched specimen is more sensitive to the effective stress due to the impurities than that in the annealed specimen within the temperature.
5. Suitable force-distance relation varies by the annealing treatment, and then it makes a difference in activation energy between the two kinds of specimens: the quenched specimen and the annealed specimen. The two values of $\Delta H(T_c)$=0.43 eV and ΔG_0=0.26 eV are smaller for the dislocation motion in the annealed specimen, as compared with $\Delta H(T_c)$=0.79 eV and ΔG_0=0.39 eV for the quenched specimen.

References

[1] Dryden, J. S.; Morimoto, S.; Cook, J. S. *Phil. Mag.* 1965, *12*, 379.
[2] Fleischer, R. L. *J. Appl. Phys.* 1962, *33*, 3504.
[3] Friedel, J. *Dislocations;* Pergamon Press; Oxford, 1964; p 224.
[4] Foreman, A. J. E.; Makin, M. J. *Phil. Mag.* 1966, *14*, 911.
[5] Conrad, H. *Acta Metall.* 1966, *14*, 1631.
[6] Ono, K. *J. Appl. Phys.* 1968, *39*, 1803.
[7] Evans, A. G.; Pratt, P. L. *Phil. Mag.* 1970, *21*, 951.
[8] Michalak, J. T. *Acta Metall.* 1965, *13*, 213.
[9] Bernstein, I. M.; Li, J. C. M.; Gensamer, M. *Acta Metall.* 1967, *15*, 801.
[10] Lee, K. S.; Park, W. K. *J. Kor. Nucl. Soc.* 1978, *10*, 73.
[11] Takeuchi, N.; Terada, K.; Yonetani, S. *Zairyo* 1978, *27*, 176 (in Japanese).
[12] Imura, T. In *Strength of Crystals*, (edited by Hasiguti, R. R.; Chikazumi, S.) Asakurashoten; Tokyo, 1968; p 28 (in Japanese).

[13] Fleischer, R. L.; Hibbard Jr., W. R. *The Relation between Structure and Mechanical Properties of Metals*, *National Physical Laboratory*, Her Majesty's Stationary Office; London, 1963; Vol. 1, pp 261-297.
[14] Johnston, W. G. *J. Appl. Phys.* 1962, *33*, 2050.
[15] Srinivasan, M.; Stoebe, T. G. *J. Mater. Sci.* 1974, *9*, 121.
[16] Hart, S. *Brit. J. Appl. Phys.* (*J. Phys. D*) ser. 2, 1968, *1*, 1285.
[17] Sprackling, M. T. *The Plastic Deformation of Simple Ionic Crystals*, (edited by Alper, A. M.; Margrave, J. L.; Nowick, A. S.) Academic Press; London, 1976; p 141.
[18] Cook, J. S.; Dryden, J. S. *Proc. Phys. Soc.* 1962, *80*, 479.
[19] Sumino, K. *Jpn. Inst. Met.* 1971, *10*, 758 (in Japanese).

Chapter 8

Effective Stress and Critical Temperature of KCl Doped with Monovalent Cations

8.1. Introduction

From the strain-rate cycling tests under application of ultrasonic oscillatory stress during plastic deformation, previous chapters have been presenting the relative curve of strain-rate sensitivity and stress decrement and information on the interaction between a mobile dislocation and various divalent impurities in KCl doped with Mg^{2+} (0.035 mol.% in the melt), Ca^{2+} (0.035 and 0.065 mol.% in the melt), Sr^{2+} (0.035, 0.050 and 0.065 mol.% in the melt) or Ba^{2+} (0.050 and 0.065 mol.% in the melt) as divalent impurities. The strain-rate cycling tests associated with the ultrasonic oscillation can separate the contributions arising from the interaction between a dislocation and impurities and that between the dislocations themselves during plastic deformation. The relative curve of strain-rate sensitivity and stress decrement reflects the influence of ultrasonic oscillation on the dislocation motion on the slip plane containing many impurities and a few forest dislocations. In this chapter, various force-distance relations which represent the interaction between a dislocation and an impurity are described with a view to selecting the most suitable one from the information given by the above-mentioned method, which was detailed in chapter 2, for KCl doped with monovalent cationic impurities (Li^+ (0.50 mol.% in the melt) or Na^+ (0.50 mol.% in the melt)). Furthermore, a critical temperature and the activation energy for the interaction between a dislocation and the impurity are obtained for various force-distance relations.

8.2. Relation between the Strain-Rate Sensitivity and the Stress Decrement

Figure 8-1 shows the variation of strain-rate sensitivity ($\Delta\tau'/\Delta\ln\dot{\varepsilon}$) and stress decrement ($\Delta\tau$) with shear strain ($\varepsilon$) for KCl:$Na^+$ (0.50 mol.% in the melt) at 180 K. The $\Delta\tau$ does not change significantly with shear strain but increases with stress amplitude at a given temperature and shear strain. The strain-rate sensitivity increases with shear strain and decreases with stress amplitude. These phenomena such as figure 8-1 are similar to the case of KCl doped with divalent impurities (see sec. 2.4 of chap. 2).

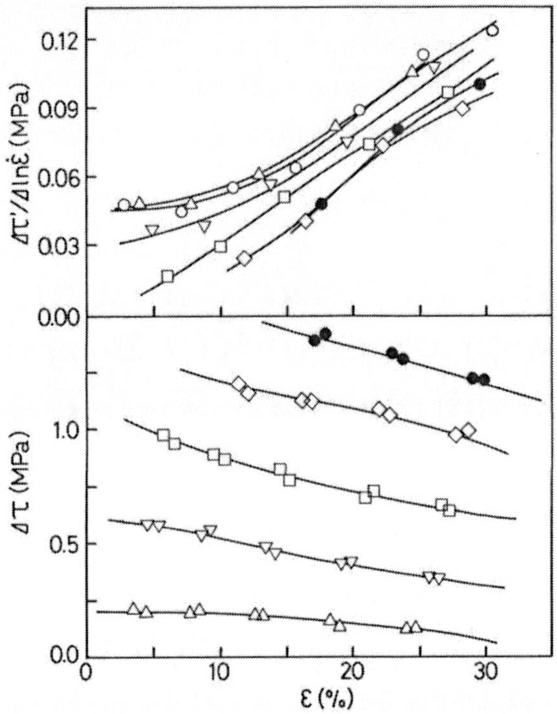

Figure 8-1. Variation of the strain-rate sensitivity ($\Delta\tau'/\Delta\ln\dot{\varepsilon}$) of flow stress and the stress decrement ($\Delta\tau$) with shear strain at 180 K and various stress amplitude for KCl:Na$^+$ (0.50 mol.% in the melt). τ_v (arb. units): (○) 0, (△) 4, (▽) 11, (□) 20, (◇) 30, (●) 37.

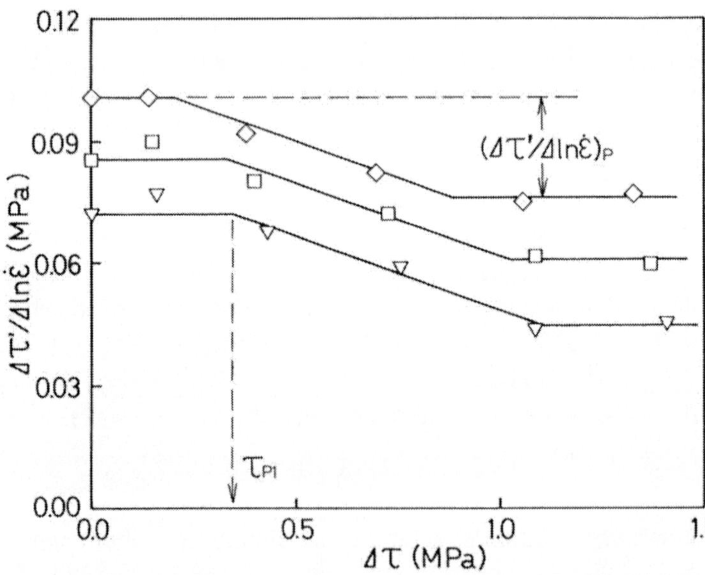

Figure 8-2. Relation between the strain-rate sensitivity and the stress decrement for KCl:Na$^+$ (0.50 mol.% in the melt) at 180 K and strains: ε = (▽) 17 %, (□) 20 %, and (◇) 23 %.

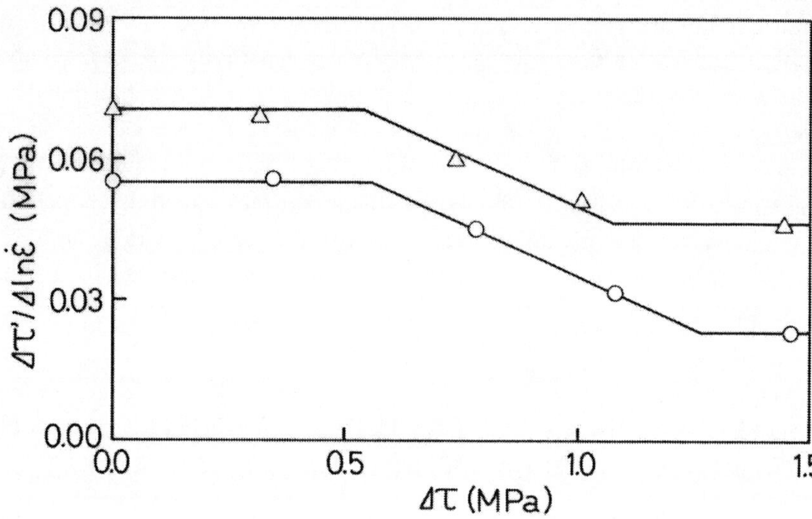

Figure 8-3. Relation between the strain-rate sensitivity and the stress decrement for KCl:Na^+ (0.50 mol.% in the melt) at 147 K and strains: ε = (○) 10 %, (△) 14 %.

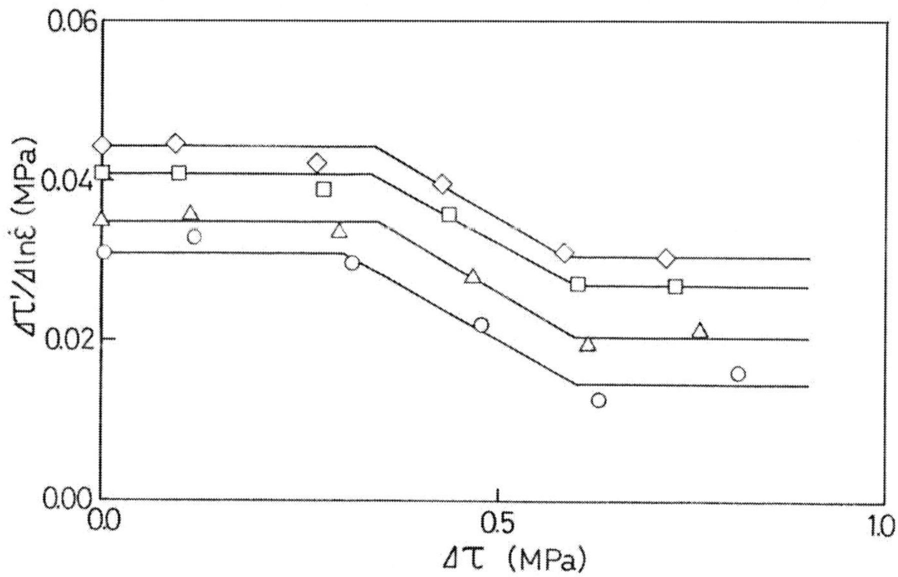

Figure 8-4. Relation between the strain-rate sensitivity and the stress decrement for KCl:Li^+ (0.50 mol.% in the melt) at 139 K and various shear strains: ε = (○) 6 %, (△) 8 %, (□) 10 %, and (◇) 11 %.

The values of strain-rate sensitivity and $\Delta\tau$ obtained from figure 8-1 at shear strains of 17, 20, and 23 % are plotted in figure 8-2, which provides the relationship between strain-rate sensitivity and $\Delta\tau$ for a fixed internal structure of the KCl:Na^+ (0.50 mol.% in the melt) single crystal at 180 K. Figure 8-3 concerns KCl:Na^+ (0.50 mol.% in the melt) single crystal at the lower temperature of 147 K. Figure 8-4 shows the relative curve of strain-rate sensitivity and stress decrement at several strains for KCl:Li^+ (0.50 mol.% in the melt) at 139 K. The strain-rate sensitivity varies with stress decrement at a given shear strain. Each curve in figures 8-2 to 8-4 has two bending points and two plateau regions. The first plateau place ranges below the first bending point at low stress decrement and the second one extends from the second bending point at high stress decrement. The strain-rate sensitivity decreases with the stress decrement between the two bending points. The value of $\Delta\tau$ at the first bending point is referred to as τ_{p1}. Figures 8-2 and 8-4 show the influence of shear strain on the relationship between strain-rate sensitivity and stress decrement. The curve shifts upward with increasing shear strain. This will be caused by the part of the strain-rate sensitivity which depends on dislocation cuttings, because the strain-rate sensitivity due to dislocation cuttings increases with strain. As can be seen from figures 8-2 and 8-3, τ_{p1} decreases when the temperature increases from 147 to 180 K.

8.3 Influence of Temperature on the Strain-Rate Sensitivity Versus Stress Decrement Curve

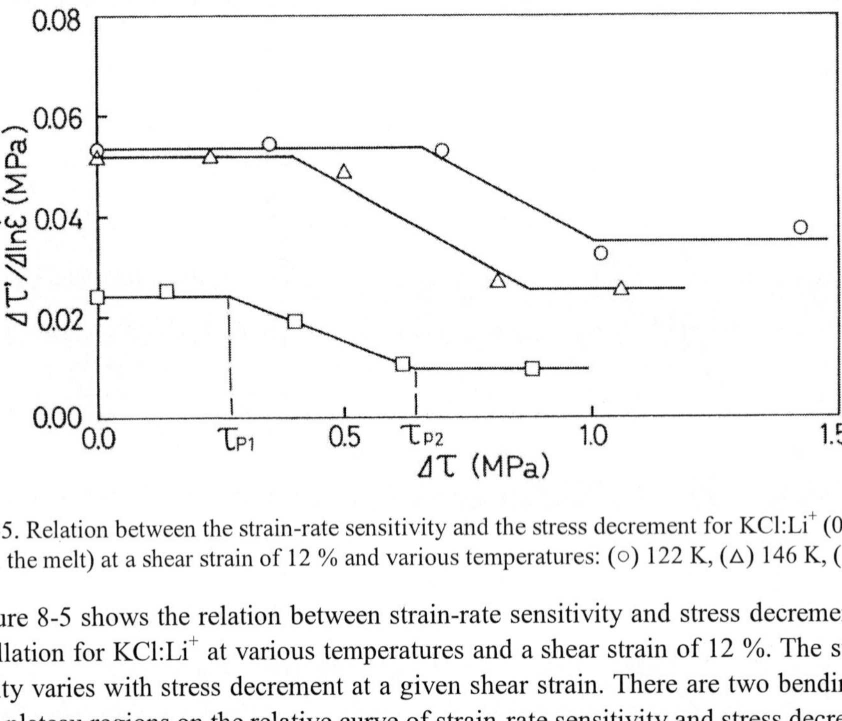

Figure 8-5. Relation between the strain-rate sensitivity and the stress decrement for $KCl:Li^+$ (0.50 mol.% in the melt) at a shear strain of 12 % and various temperatures: (○) 122 K, (△) 146 K, (□) 177 K.

Figure 8-5 shows the relation between strain-rate sensitivity and stress decrement due to the oscillation for $KCl:Li^+$ at various temperatures and a shear strain of 12 %. The strain-rate sensitivity varies with stress decrement at a given shear strain. There are two bending points and two plateau regions on the relative curve of strain-rate sensitivity and stress decrement for KCl doped with Li^+, as shown in figure 8-5. The first bending point at low stress decrement has been denoted by τ_{p1} and the second one at high stress decrement by τ_{p2}. As the temperature increases, τ_{p1} and τ_{p2} shift in the direction of low stress decrement. The same phenomenon as figure 8-5 is also observed for $KCl:Na^+$. τ_{p1}, however, does not appear at a temperature above 221 K for $KCl:Li^+$ and at above 279 K for $KCl:Na^+$. The strain-rate sensitivity versus stress decrement curve for $KCl:Li^+$ at 299 K is shown in figure 8-6 (a) and that for $KCl:Na^+$ at 279 K in figure 8-6 (b). For both the specimens, τ_{p1} does not appear on the relation curves of strain-rate sensitivity and stress decrement at various shear strains. These phenomena can be understood as follows. An application of ultrasonic oscillatory stress with low amplitude cannot influence the average length of the dislocation segment at low temperature as shown in figure 8-5, but even the stress with low amplitude can do so at such a temperature as figure 8-6 (a) for $KCl:Li^+$ and (b) for $KCl:Na^+$, because τ_{p1} is considered to represent the effective stress due to only one type of the impurities which lie on the mobile dislocation when the dislocation moves forward with the help of oscillation (see chap. 2).

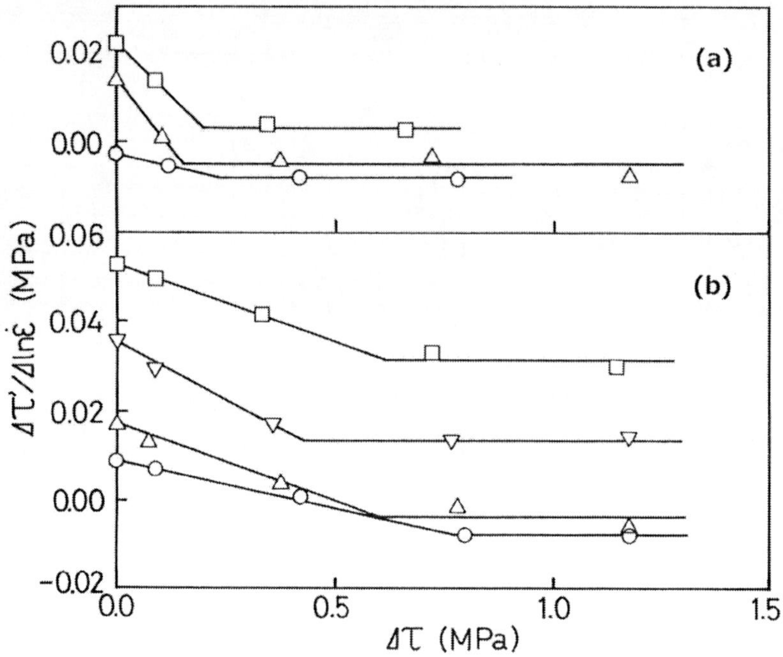

Figure 8-6. Relation between the strain-rate sensitivity and the stress decrement for (a) $KCl:Li^+$ (0.50 mol.% in the melt) at 299 K and shear strains: ε = (○) 24 %, (△) 28 %, (□) 32 % and (b) $KCl:Na^+$ (0.50 mol.% in the melt) at 279 K and shear strains: ε = (○) 8 %, (△) 14 %, (▽) 20 %, (□) 26 %.

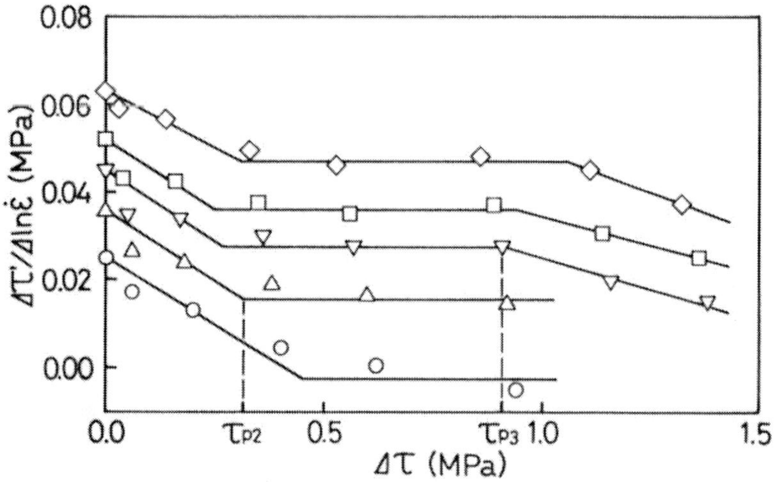

Figure 8-7. Relation between the strain-rate sensitivity and the stress decrement for $KCl:Li^+$ (0.50 mol.% in the melt) at 249 K and various shear strains: ε = (○) 25 %, (△) 30 %, (▽) 33 %, (□) 35 %, and (◇) 38 %.

The strain-rate sensitivity becomes negative at shear strains of 24 %, 25 % ($\Delta\tau > 0.41$ MPa), and 28 % ($\Delta\tau > 0.11$ MPa) for KCl:Li^+ and at shear strains of 8 % ($\Delta\tau > 0.42$ MPa) and 14 % ($\Delta\tau > 0.50$ MPa) for KCl:Na^+ as shown in figures 8-6 (a), (b), and 8-7. The negative strain-rate sensitivity appears on the relative curves of strain-rate sensitivity and stress decrement above 216 K for KCl:Li^+ and above 246 K for KCl:Na^+.

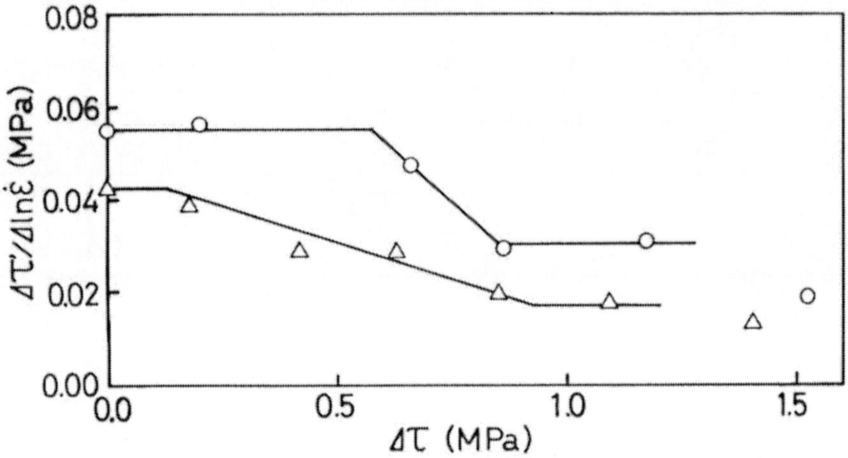

Figure 8-8. Relation between the strain-rate sensitivity and the stress decrement for KCl:Li^+ (0.50 mol.% in the melt) at various conditions: (○) 160 K and ε =17 %, (△) 192 K and ε =15 %.

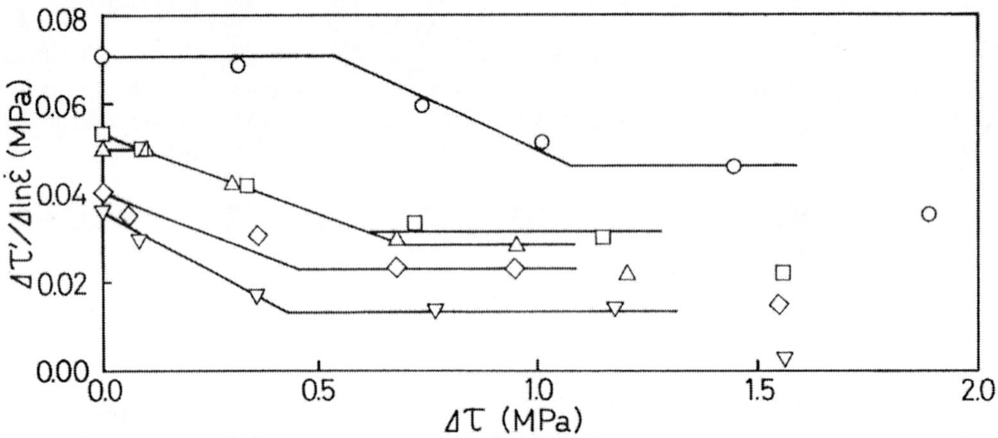

Figure 8-9. Relation between the strain-rate sensitivity and the stress decrement for KCl:Na^+ (0.50 mol.% in the melt) at various conditions: (○) 147 K and ε =14 %, (△) 251 K and ε =19 %, (▽) 279 K and ε =20 %, (□) 279 K and ε =26 %, (◇) 300 K and ε =28 %.

A third bending point is found on the relative curves of strain-rate sensitivity and stress decrement for KCl:Li^+ at shear strains of 33, 35, and 38 % as shown in figure 8-7. The third bending point is denoted by τ_{p3}. However, τ_{p3} does not appear on the curves at shear strains of 25 and 30 %, because high stress amplitude could not be applied to the specimen during the compression test. The lengths of second plateau regions on the curves at shear strains of 33, 35 and 38 % are 0.64, 0.69 and 0.75 MPa, respectively. The second plateau region where the obstacles to the dislocation motion are considered to be only strong ones such as forest dislocations (see sec. 2.7 of chap. 2) becomes longer with increasing shear strain. The appearance of existence of τ_{p3} is also observed on the relative curves of strain-rate sensitivity and stress decrement at 160 and 192 K for KCl:Li$^+$ in figure 8-8 and at 147, 251, 279 and 300 K for KCl:Na$^+$ in figure 8-9.

8.4. Dependence of τ_{P1}, τ_{P2}, and τ_Y on the Temperature

The dependence of τ_{p1}, τ_{p2}, and yield stress (critical resolved shear stress (τ_y)) on temperature is shown in figure 8-10 (a) for KCl:Li$^+$ and figure 8-10 (b) for KCl:Na$^+$. The plots of τ_{p1} and τ_{p2} in figures 8-10 (a) and (b) are the mean value at a given temperature because τ_{p1} and τ_{p2} vary slightly with shear strain. The τ_y is determined at the intersection of the tangent to the easy glide region at a stress-strain curve and the straight line extrapolated from the elastic region at the stress-strain curve. The curve of τ_{p1} is fitted to the data by using the least squares method, assuming that the force-distance profile between the dislocation and the impurity is a square [1] and the two curves of τ_{p2} and τ_y are to guide the reader's eye. τ_{p1} and τ_{p2} decrease with increasing temperature as the variation of τ_y within the temperature for both the specimens. At high temperature, the critical temperature, T_c, at which τ_{p1} is zero can be estimated to be around 250 K for KCl:Li$^+$ and 270 K for KCl:Na$^+$ from figures 8-10 (a) and (b). Determination of T_c has been attempted from the temperature dependence of yield stress so far [2-7]. Then, the effective stress due to impurities was defined by subtraction of the almost constant yield stress (this region is called "plateau") at high temperature such as room temperature from the yield stress at a given low temperature. That is, the yield stress at high temperature was taken as the internal stress for the specimen. However, the value of internal stress in this assumption may be doubtful because the yield stress seems to be gradually decreased over the temperature of T_c. The same can be also observed in figures 8-10 (a) and (b). This is because an athermal component of yield stress depends on temperature only through the shear modulus. Further, it is difficult to investigate the interaction between a dislocation and weak obstacles during plastic deformation from yield stress by reason that yield stress depends on dislocation velocity, dislocation density, and multiplication of dislocations [8]. τ_{p2} for KCl:Li$^+$ and KCl:Na$^+$ increases with decreasing temperature and gradually approaches the τ_{p1} curve, which is the same phenomena for KCl:Br$^-$ and KCl:I$^-$ [9], as shown in figures 8-10 (a) and (b).

8.5. Interpretation of Strain-Rate Sensitivity Versus Stress Decrement Curve

We consider a dislocation that is pinned by many weak obstacles and is bowed by the applied stress between a few strong obstacles, such as forest dislocations, during plastic deformation. The weak obstacles are considered to be the impurities and not to be vacancies in this study, because the vacancies in the specimen have low density in contrast to the impurities. For example, the concentration of vacancies in thermal equilibrium state is about 1.2×10^{-7} % for a KCl single crystal quenched from 673 K to room temperature.

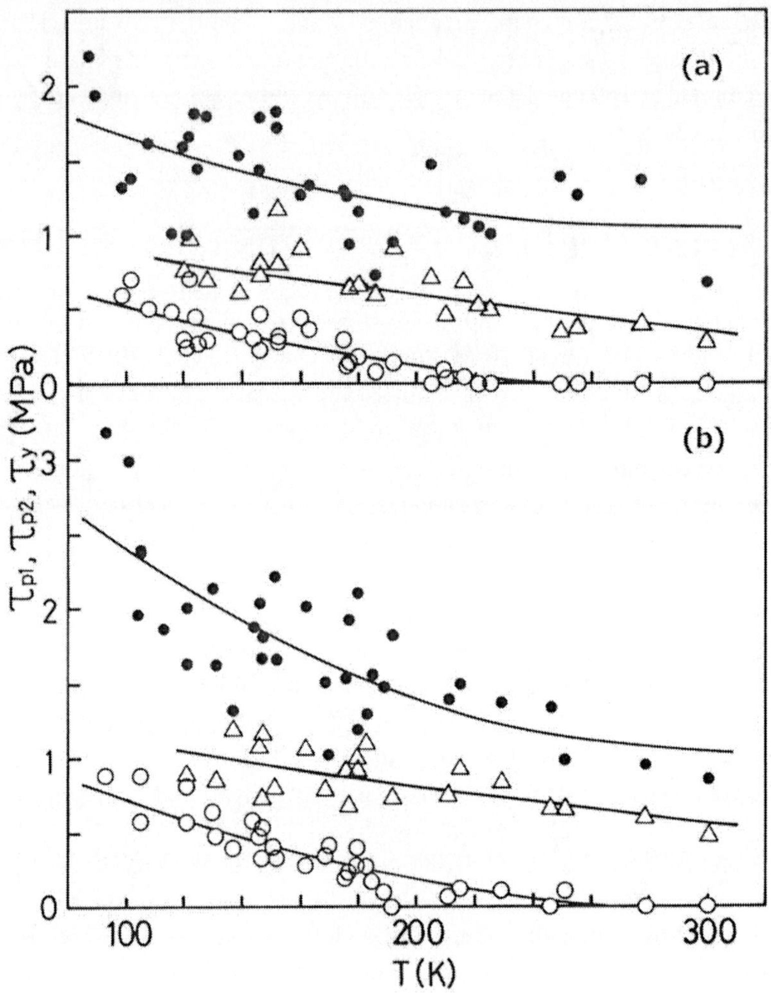

Figure 8-10. Dependence of (○) τ_{p1}, (△) τ_{p2}, and (•) τ_y on the temperature for (a) KCl:Li^+ (0.50 mol.% in the melt) and (b) KCl:Na^+ (0.50 mol.% in the melt).

The strain-rate sensitivity of flow stress is inversely proportional to the activation volume, i.e. the average length of dislocation segments. The average length of dislocation segments depends on the distribution of impurities on mobile dislocation. The strain-rate sensitivity obtained in this study reflects the effect of ultrasonic oscillation on the distribution of them. Therefore, the first plateau region in figures 8-2 to 8-5 indicates that the distribution of impurities on a mobile dislocation remains constant there. That is to say, all impurities act as obstacles to the dislocation motion, and the average length of dislocation segments is considered to remain constant in the region. When the stress amplitude increases, the dislocation begins to break away from the impurities by oscillation between the forest dislocations and the average length of the dislocation begins to increase. Then, the strain-rate sensitivity starts to decrease at the first bending point, τ_{p1}. The distribution of impurities on a mobile dislocation begins to change due to the oscillation and some impurities stop acting as obstacles there. The phenomenon that the oscillating dislocation increases the average length of its segment is observed in internal friction measurements [10]. Furthermore, it has been reported that the length of the dislocation segment increases and the strain-rate sensitivity decreases when the ultrasonic oscillatory stress is applied to the alkali halide crystals at room temperature during plastic deformation [11]. Application of oscillations with low stress amplitude cannot influence the average length of dislocation segments at low temperature, but it can do so at high temperature. This is shown in figures 8-2, 8-3, 8-5, and 8-6. A similar experiment was carried out for nominally pure KCl single crystals (see sec. 2.4 of chap. 2). As a result, the first bending point does not appear on the strain-rate sensitivity versus stress decrement curve. As for KCl single crystals doped with various impurities, no bending point appears on the curve (see sec. 2.4 of chap. 2). The experimental fact that the relative curve of strain-rate sensitivity and the stress decrement has two bending points and two plateau regions is observed only for crystals doped with one type of impurity. Therefore, τ_{p1} is considered to be due to additive ions (Li^+ and Na^+ in this chapter). Ohgaku and Takeuchi [9] previously conducted a similar experiment for KCl single crystals doped with monovalent impurities (Br^- or I^-) and found the phenomenon that the plots of the strain-rate sensitivity and the stress decrement due to application of oscillation have two bending points and two plateau regions, as above. Furthermore, it was reported that the first bending point (τ_{p1}) increases proportionally to the yield stress; that is to say, τ_{p1} increases depending on the impurity concentration, because the yield stress generally increases with impurity concentration [2-6]. They also described how the temperature and impurity concentration dependence of τ_{p1} suggests that the phenomenon is caused by the interaction between a dislocation and weak obstacles such as Br^- or I^-. The value of τ_{p1}, accordingly, is regarded as the effective stress due to the impurities (weak obstacles) that lie on the mobile dislocation when the dislocation moves with the aid of thermal activation.

Increasing the stress amplitude further, the dislocation breaks away from all impurities and the strain-rate sensitivity becomes constant. The obstacles to the dislocation motion are only forest dislocations and the impurities no longer act as obstacles. Then, the second plateau region appears on the curve. Consequently, the strain-rate sensitivity versus stress decrement curve reflects the effect of ultrasonic oscillation on the dislocation motion on the slip plane containing many impurities and a few forest dislocations, during plastic deformation.

8.6. Relation between the Effective Stress Due to Impurities and Temperature

The critical temperature (T_c) at which the relative curve of τ_{p1} and temperature intersects the abscissa and τ_{p1} is zero, may be determined to be around 240 K for $KCl{:}Li^+$ and 260 K for $KCl{:}Na^+$ from figures 8-10 (a) and (b). τ_y seems to approach an almost constant value above T_c. Determination of T_c has been attempted from the temperature dependence of the yield stress so far [2-7]. Then, the effective stress due to impurities was defined by subtraction of the almost constant yield stress at high temperature (such as room temperature) from the yield stress at a given low temperature. That is, the yield stress at high temperature was taken as the internal stress for the specimen. However, the value of internal stress in this assumption may be doubtful, since the yield stress seems to be gradually decreased over the temperature of T_c. The same can be also observed in figures 8-10 (a) and (b). This is because an athermal component of yield stress depends on temperature only through the shear modulus. Further, it is difficult to investigate the interaction between a dislocation and weak obstacles during plastic deformation from the yield stress for the reason that yield stress depends on dislocation velocity, dislocation density, and multiplication of dislocations [8].

In this section, the three models [1], which are a square force-distance relation, a parabolic one, and a triangular one, taking account of the Friedel relation [12] are examined on the basis of the relation between τ_{p1} and temperature for both the specimens. The square force-distance relation, the parabolic one, and the triangular one are termed the SQ, the PA, and the TR, respectively in this chapter. The value of τ_{p1} should depend on the type of weak obstacles. Observation of τ_{p1}, therefore, has provided information on the interaction between a dislocation and weak obstacles to dislocation motion (see chap. 2). That is, the temperature dependence of τ_{p1} reveals the force-distance profile, which expresses the interaction between a dislocation and a weak obstacle. When alkali halide crystals are doped with monovalent impurities, cubic symmetry defects are produced around the impurities. Then, the interaction between a dislocation and the impurity is expected to cause gradual hardening (see chap. 1). Figures 8-11 (a) and (b) show that the interaction between a dislocation and the impurity for KCl:Li^+ and KCl:Na^+ can be approximated to the SQ of which the linear relationship of τ_{p1} and temperature, T, is given by $\tau_{p1}^{2/3}$ versus T. The slopes of straight lines in the figure are determined by the method of least squares. The critical temperatures (T_c) at which the lines intersect the abscissa in figures 8-11 (a) and (b) and τ_{p1} is zero, are determined to be 254 K for KCl:Li^+ and 267 K for KCl:Na^+. T_c is not influenced by the impurity concentration (see sec. 2.5 of chap. 2 or ref. [9]), whereas it depends on strain rate $\dot{\varepsilon}$ in accordance with

$$T_c = \frac{\Delta G_0}{k} \ln\left\{\frac{\dot{\varepsilon}}{\rho b^2 \nu_D (L_0/L)^2}\right\}, \tag{8-1}$$

where ΔG_0 is the Gibbs free energy for the breakaway of the dislocation from an impurity in the absence of an applied stress, k is the Boltzmann constant, ρ is the density of mobile dislocations, b is the magnitude of the Burgers vector, and ν_D is the Debye frequency. τ_{p0}, which is obtained by extrapolating the line to 0 K, is 1.10 MPa for KCl:Li^+ (0.50 mol.% in the melt) and 1.45 MPa for KCl:Na^+ (0.50 mol.% in the melt). τ_{p0} is the effective stress due to the impurities without thermal activation. If the interaction between a dislocation and the impurity for both the specimens can be approximated to another two models, that is, the PA or the TR, T_c and τ_{p0} are given in table 8-1. On the three force-distance relations, the value of T_c is almost the same, irrespective of the kind of force-distance relation for both the specimens, and the value of T_c for KCl:Li^+ is obviously small in comparison with that for KCl:Na^+ from table 8-1. At the temperature of T_c, thermal fluctuations can provide the entire energy for breaking through the impurity. It was previously found that when the divalent cationic size becomes closer to the K^+ ionic size, T_c tends to increase for KCl doped with Mg^{2+}, Ca^{2+}, Sr^{2+} or Ba^{2+} (see sec. 4.3 of chap. 4). It may also be predicted for KCl doped with monovalent impurities (cations) that when the size of monovalent impurity is increasingly close to that of the K^+ from the small monovalent cationic size side, T_c tends to increase.

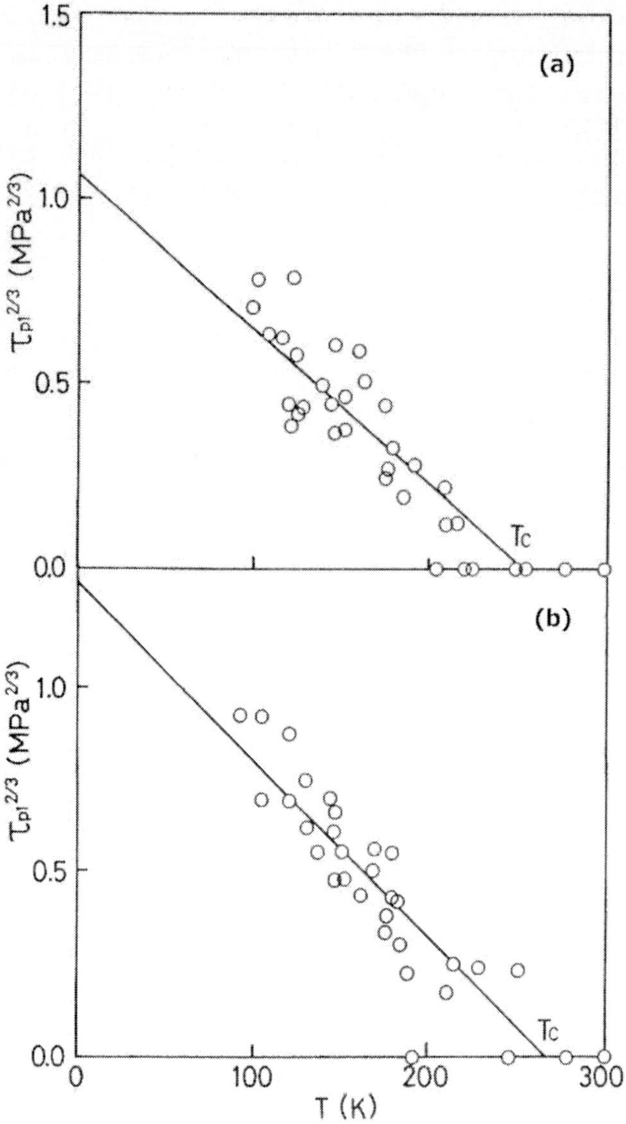

Figure 8-11. Linear plots of $\tau_{p1}^{2/3}$ and T for (a) KCl:Li^{+} (0.50 mol.% in the melt) and (b) KCl:Na^{+} (0.50 mol.% in the melt).

Table 8-1. Values of T_c and τ_{p0} for the three force-distance relations between a dislocation and the impurity

The force-distance relation	Single crystal[a]	T_c (K)	τ_{p0} (MPa)
Square	KCl:Li^{+}	254	1.10
	KCl:Na^{+}	267	1.45
Parabola	KCl:Li^{+}	256	1.73
	KCl:Na^{+}	266	2.29
Triangle	KCl:Li^{+}	257	2.44
	KCl:Na^{+}	271	3.16

a The concentration of Li+ or Na+ in KCl is 0.50 mol.% in the melt.

The separation of various models between a dislocation and an impurity for a particular case has been conducted from the effect of the impurity concentration on the flow stress at 0 K and the variation of the thermal component of the flow stress with temperature (see sec. 1.2 of chap. 1 or ref. [13]). On the basis of the relation between τ_{p1} and temperature, the most suitable model is investigated here. The SQ gives the following relation

$$\left(\frac{\tau_{p1}}{\tau_{p0}}\right)^{2/3} = 1 - \left(\frac{T}{T_c}\right), \tag{8-2}$$

the PA gives

$$\left(\frac{\tau_{p1}}{\tau_{p0}}\right)^{2/3} = 1 - \left(\frac{T}{T_c}\right)^{2/3}, \tag{8-3}$$

and the TR gives

$$\left(\frac{\tau_{p1}}{\tau_{p0}}\right)^{2/3} = 1 - \left(\frac{T}{T_c}\right)^{1/2}. \tag{8-4}$$

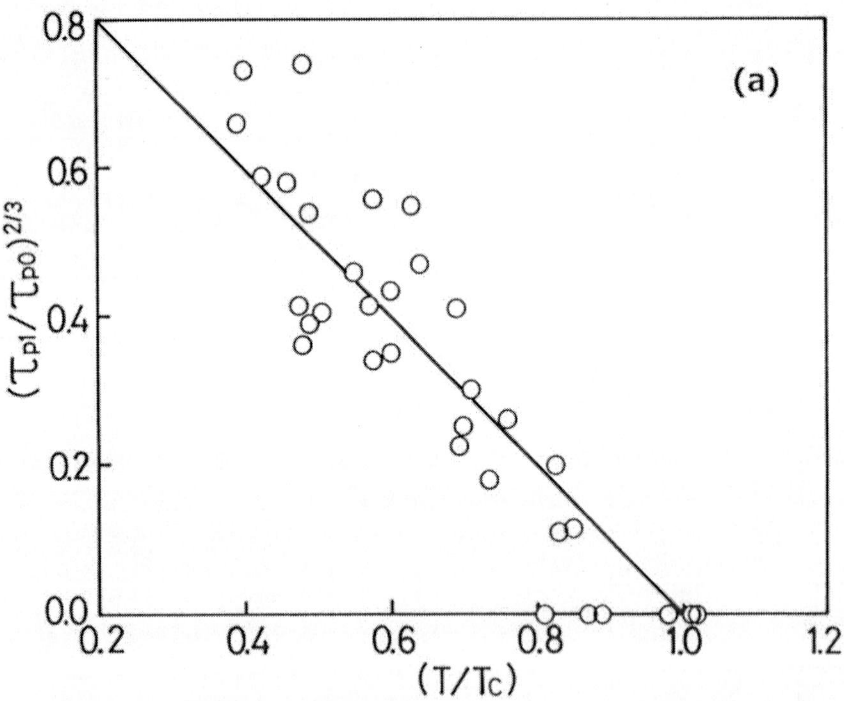

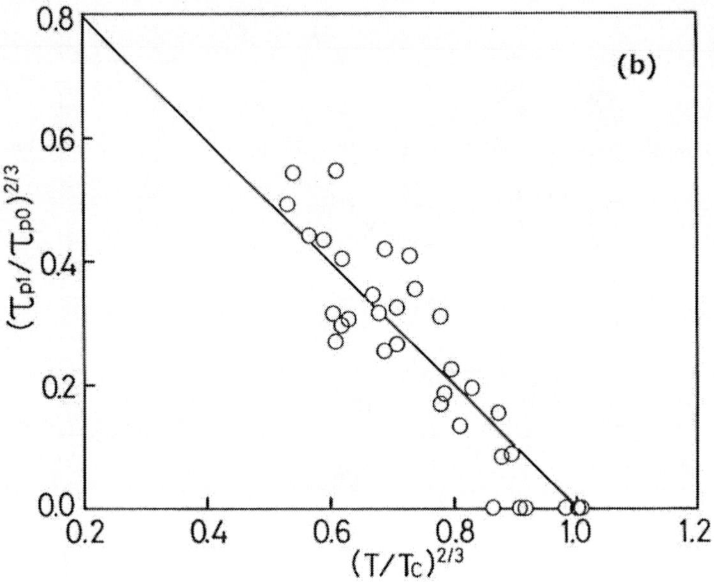

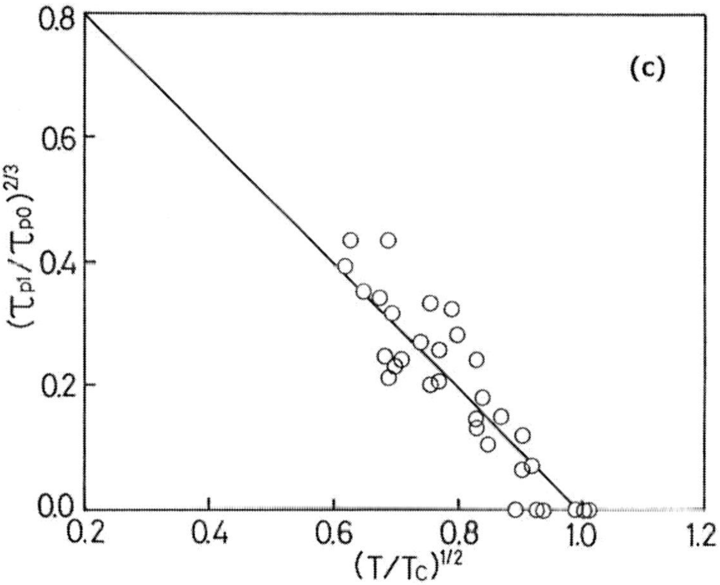

Figure 8-12. Relationship between the effective stress and the temperature for KCl:Li$^+$ (0.50 mol.% in the melt) at various models: (a) the square force-distance relation, (b) the parabolic one, and (c) the triangular one.

The results of equations (8-2) to (8-4) are shown in figures 8-12 (a) to (c) for KCl:Li$^+$ (0.50 mol.% in the melt) and figures 8-13 (a) to (c) for KCl:Na$^+$ (0.50 mol.% in the melt). The slopes of straight lines in figures 8-12 and 8-13 are determined by the method of least squares. However, as can be seen from figures 8-12 and 8-13, there is no difference in the linear effective stress-temperature relation for the three models as Ono's description [14]: it is impossible to select one of several models on the basis of an experimentally obtained relationship of applied stress and temperature. With regard to LiF:Mg^{2+} (150 ppm), it has been reported that the observation of a linear relationship in various diagrams of flow stress and temperature is not a sufficient criterion to determine the hardening mechanism [15].

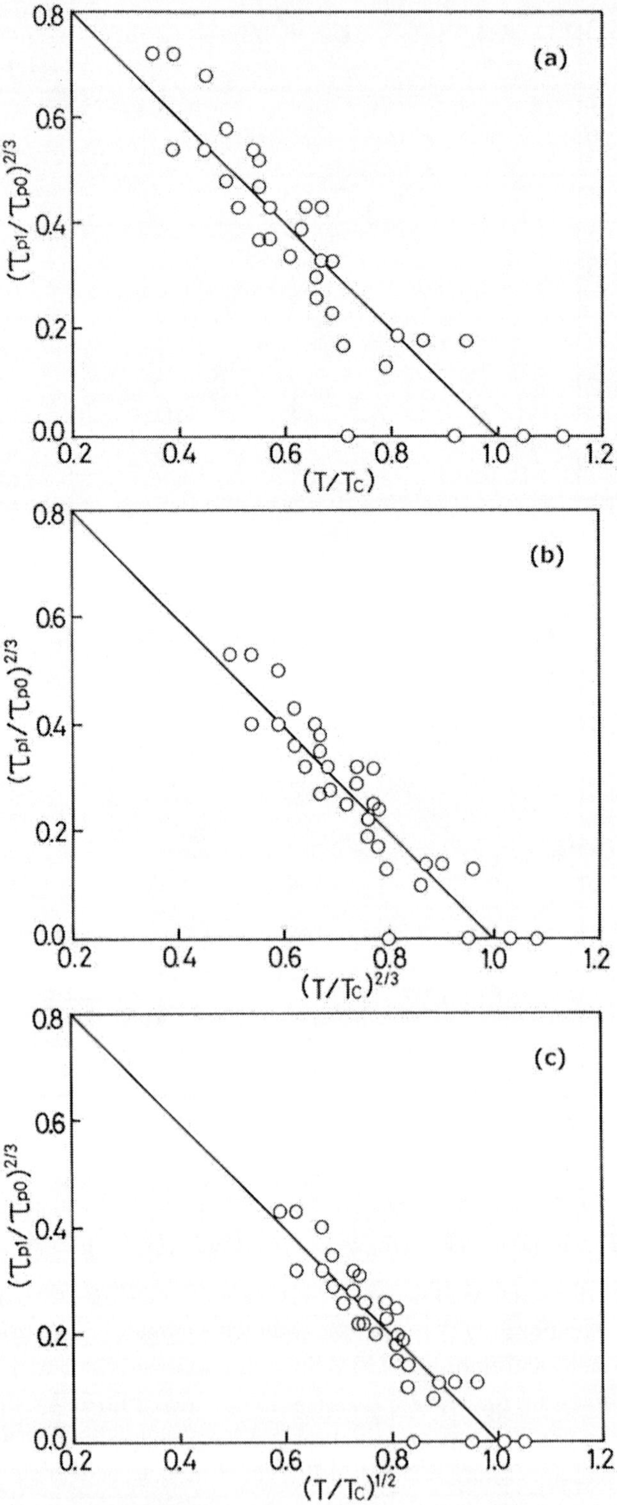

Figure 8-13. Relationship between the effective stress and the temperature for KCl:Na^+ (0.50 mol.% in the melt) at various models: (a) the square force-distance relation, (b) the parabolic one, and (c) the triangular one.

8.7. DEPENDENCE OF THE STRAIN-RATE SENSITIVITY DUE TO THE IMPURITIES ON THE TEMPERATURE

The dependence of the strain-rate sensitivity due to the impurities on the temperature was investigated as to the above-mentioned three force-distance relations [1] for both the specimens.

When the thermally activated overcoming of the substitutional impurities controls the dislocation velocity [7], the strain rate is assumed to be given by an Arrhenius-type equation [16]:

$$\dot{\varepsilon} = \dot{\varepsilon}_0 \exp\left(\frac{-\Delta G}{kT}\right), \tag{8-5}$$

where $\dot{\varepsilon}_0$ is a frequency factor, ΔG is the change in Gibbs free energy of activation for the dislocation motion. If the changes in entropy (ΔS) are neglected, ΔG of equation (8-5) may be replaced by the activation enthalpy (ΔH). That is, ΔH is expressed by [17, 18]

$$\Delta H = \alpha k T, \tag{8-6}$$

where α is a constant. ΔH can be calculated from the relation [17-20]:

$$\Delta H = -kT^2 \left(\frac{\partial \ln \dot{\varepsilon}}{\partial \tau^*}\right)_T \left(\frac{\partial \tau^*}{\partial T}\right)_{\dot{\varepsilon}}. \tag{8-7}$$

Combining equations (8-6) and (8-7), we find

$$\left(\frac{\partial \tau^*}{\partial \ln \dot{\varepsilon}}\right)_T = -\frac{T}{\alpha}\left(\frac{\partial \tau^*}{\partial T}\right)_{\dot{\varepsilon}}. \tag{8-8}$$

The $(\partial \tau^* / \partial T)_{\dot{\varepsilon}}$ of equation (8-8) can be obtained from the differentiation of equation (8-2) with respect to the temperature for the SQ as follows

$$\frac{\partial \tau_{p1}}{\partial T} = -\frac{3\tau_{p0}}{2T_c}\left(1 - \frac{T}{T_c}\right)^{1/2}. \tag{8-9}$$

In the same way, the $(\partial \tau^* / \partial T)_{\dot{\varepsilon}}$ of equation (8-8) for the PA can be obtained from equation (8-3) as follows

$$\frac{\partial \tau_{p1}}{\partial T} = -\frac{\tau_{p0}}{T_c}\left\{\left(\frac{T_c}{T}\right)^{2/3} - 1\right\}^{1/2} \tag{8-10}$$

and also from equation (8-4), that of equation (8-8) for the TR is given by

$$\frac{\partial \tau_{p1}}{\partial T} = -\frac{3\tau_{p0}}{4T_c}\left\{\left(\frac{T_c}{T}\right)-\left(\frac{T_c}{T}\right)^{1/2}\right\}^{1/2}. \tag{8-11}$$

Further substituting of equation (8-9) in equation (8-8), the strain-rate sensitivity due to the impurities for the SQ yields

$$\frac{\partial \tau^*}{\partial \ln \dot{\varepsilon}} = \frac{3\tau_{p0}T}{2\alpha T_c}\left(1-\frac{T}{T_c}\right)^{1/2}. \tag{8-12}$$

Similarly from equations (8-8) and (8-10), the strain-rate sensitivity due to the impurities for the PA is expressed by

$$\frac{\partial \tau^*}{\partial \ln \dot{\varepsilon}} = \frac{\tau_{p0}T}{\alpha T_c}\left\{\left(\frac{T_c}{T}\right)^{2/3}-1\right\}^{1/2} \tag{8-13}$$

and also from equations (8-8) and (8-11), that for the TR is expressed by

$$\frac{\partial \tau^*}{\partial \ln \dot{\varepsilon}} = \frac{3\tau_{p0}T}{4\alpha T_c}\left\{\left(\frac{T_c}{T}\right)-\left(\frac{T_c}{T}\right)^{1/2}\right\}^{1/2}. \tag{8-14}$$

Figures 8-14 (a) and (b) show the results for $KCl:Li^+$ (0.50 mol.% in the melt) and $KCl:Na^+$ (0.50 mol.% in the melt), that is, the each curve corresponds to the relation between temperature and the strain-rate sensitivity calculated from equation (8-12) for the SQ, equation (8-13) for the PA, and equation (8-14) for the TR. The relative curve of temperature and the strain-rate sensitivity due to the impurities for the SQ, that for the PA, and that for the TR are represented as a solid line, a dashed line, and a dash-dotted line, respectively. The open circles in figures 8-14 (a) and (b) correspond to the dependence of temperature and the strain-rate sensitivity due to impurities which is given by the difference between those at the first plateau place and at the second one on the relative curve of strain-rate sensitivity and stress decrement for $KCl:Li^+$ (0.50 mol.% in the melt) and $KCl:Na^+$ (0.50 mol.% in the melt) and is denoted by $(\Delta\tau'/\Delta\ln\dot{\varepsilon})_p$ in figure 8-2. The $(\Delta\tau'/\Delta\ln\dot{\varepsilon})_p$ is considered to be the strain-rate sensitivity due to the impurities. This is based on the following two experimental results. First, the $(\Delta\tau'/\Delta\ln\dot{\varepsilon})_p$ is constant independent of the strain as shown in figures 8-2 to 8-4. Second, the $(\Delta\tau'/\Delta\ln\dot{\varepsilon})_p$ is proportional to the square root of impurity concentration, which is related to the inverse of the average length of the dislocation segment, at a given temperature (see sec. 4.4 of chap. 4). On the hypothesis of the break-away type by Granato and Lücke [21, 22], we would additionally expect so from the result of $\ln Q_H^{-1}$ versus $\ln\Delta\tau$ as described by Ohgaku and Takeuchi [23, 24]: the brief outline is given in the "Introduction section" of chapter 2. Q_H^{-1} is the amplitude dependent region of internal friction.

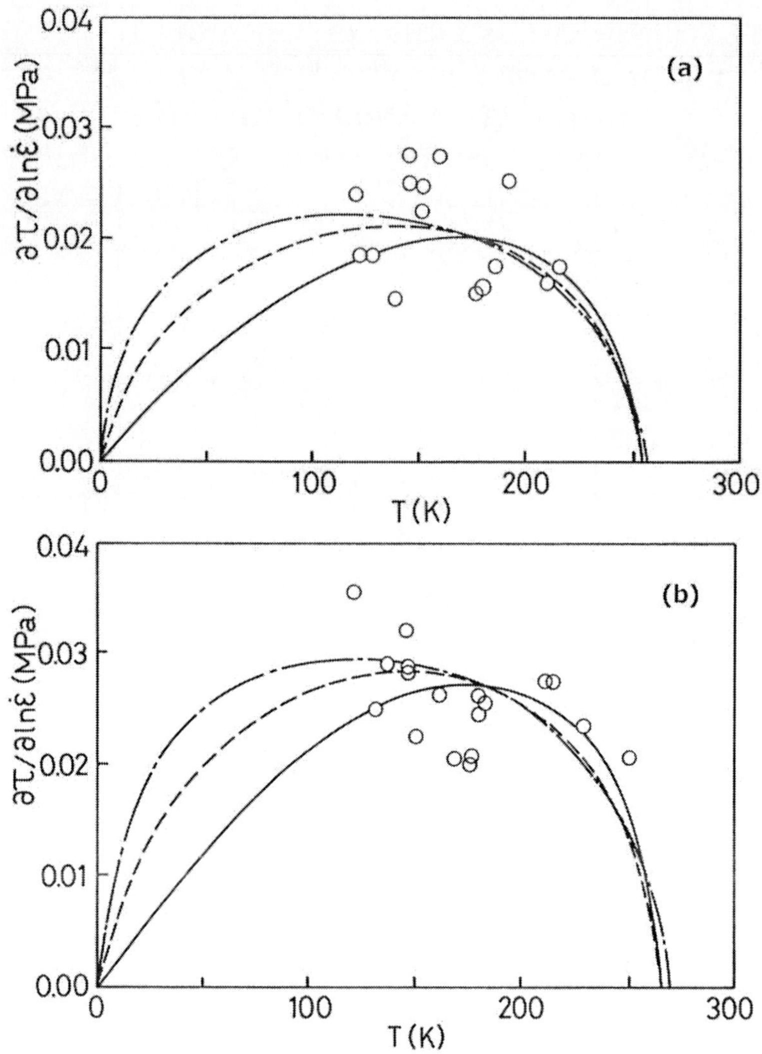

Figure 8-14. Dependence of the strain-rate sensitivity due to the impurities on the temperature for (a) $KCl:Li^+$ (0.50 mol.% in the melt) and (b) $KCl:Na^+$ (0.50 mol.% in the melt). (○) $(\Delta\tau'/\Delta\ln\dot{\varepsilon})_p$ is predicted by various models: (—) the square force-distance relation, (- - -) the parabolic one, and (—·—) the triangular one.

As can be seen from figures 8-14 (a) and (b), it is difficult to determine which force-distance relation is appropriate to the relation between temperature and $(\Delta\tau'/\Delta\ln\dot{\varepsilon})_p$ for both the specimens within the present temperatures. However, if the relation between temperature and $(\Delta\tau'/\Delta\ln\dot{\varepsilon})_p$ can be obtained below approximate 100 K, it will be possible to select the most suitable force-distance relation of the three for both the specimens on the basis of the dependence of the strain-rate sensitivity due to the impurities on the temperature. This is because there is a great distinction in the strain-rate sensitivity due to the impurities for the three force-distance relations at the constant temperature below 100 K in comparison with that above 100 K, as shown in figures 8-14 (a) and (b).

8.8. Relation between Temperature and the Activation Enthalpy for the Interaction between a Dislocation and the Impurity

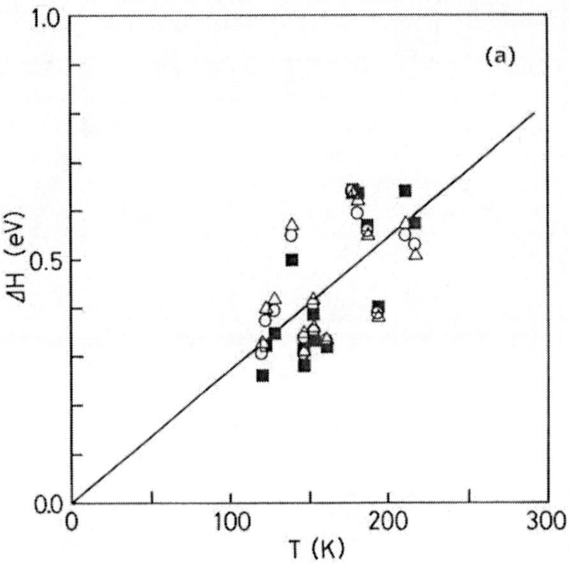

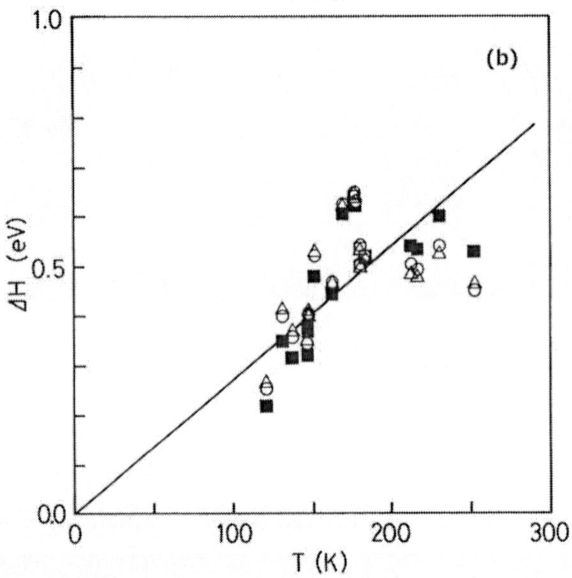

Figure 8-15. Proportional relationship between the temperature and the activation enthalpy for the interaction between a dislocation and the impurity for (a) KCl:Li^+ (0.50 mol.% in the melt) and (b) KCl:Na^+ (0.50 mol.% in the melt) at various models: (■) the square force-distance relation, (○) the parabolic one, and (Δ) the triangular one.

The relation between temperature and the activation enthalpy for the interaction between a dislocation and the impurity was investigated for both the specimens about the above-mentioned three force-distance relations [1]. The activation enthalpy for the interaction between a dislocation and the impurity can be obtained from the substitution of equation (8-9) for the SQ, equation (8-10) for the PA, and equation (8-11) for the TR in equation (8-7). As for $(\partial \ln \dot{\varepsilon} / \partial \tau)_T$ in equation (8-7), that is replaced by $(\Delta \ln \dot{\varepsilon} / \Delta \tau')_p$ as a reciprocal of the strain-rate sensitivity due to impurities. The results of calculations for ΔH against the temperature, $\Delta H(T)$, at the three force-distance relations are shown in figure 8-15 (a) for $KCl{:}Li^+$ and (b) for $KCl{:}Na^+$. The slope of $\Delta H(T)$, which is determined by the method of least squares, is quite coincidental for the three force-distance relations for both the specimens. From the $\Delta H(T)$, the activation enthalpy for the SQ seems to be nearly proportional to the temperature in comparison with the cases of the PA and the TR for $KCl{:}Na^+$ as shown in figure 8-15 (b). However, it is difficult to select the most suitable model of the three for $KCl{:}Li^+$ from the proportionality of $\Delta H(T)$ in figure 8-15 (a). The value of $\Delta H(T_c)$ taken from the relation between temperature and the activation enthalpy for the interaction between a dislocation and the impurity is given in table 8-2. The value of $\Delta H(T_c)$ corresponds to the activation enthalpy for overcoming the strain field with cubic distortion around the substitutional impurity by a dislocation without an applied stress. On the three force-distance relations, the value of $\Delta H(T_c)$ is almost the same, irrespective of the kind of force-distance relation for both the specimens and the value of $\Delta H(T_c)$ for $KCl{:}Li^+$ is nearly equal to that for $KCl{:}Na^+$ from table 8-2.

Table 8-2. Values of $\Delta H(T_C)$ for the three force-distance relations between a dislocation and the impurity

The force-distance relation	Single crystal	$\Delta H(T_C)$ (eV)
Square	$KCl{:}Li^+$	0.69
	$KCl{:}Na^+$	0.72
Parabola	$KCl{:}Li^+$	0.70
	$KCl{:}Na^+$	0.72
Triangle	$KCl{:}Li^+$	0.71
	$KCl{:}Na^+$	0.73

8.9. Gibbs Free Energy for the Breakaway of a Dislocation from the Impurity

The Gibbs free energy for the square force-distance profile between a dislocation and the impurity is given by [25]

$$\Delta G = \Delta G_0 - \beta \tau^{*2/3} \quad (\beta = (2\mu b^4 d^3 L_0^2)^{1/3}), \tag{8-15}$$

where μ is the shear modulus and d is the activation distance. Substituting equation (8-15) into equation (8-5) and differentiating with respect to the effective shear stress (τ^*), we find

$$\tau^* \frac{\partial \ln \dot{\varepsilon}}{\partial \tau^*} = \frac{2G_0}{3kT} + \frac{2}{3}\ln(\dot{\varepsilon}/\dot{\varepsilon}_0). \tag{8-16}$$

Equation (8-16) will be transformed to

$$\tau_{p1}(\Delta \ln \dot{\varepsilon}/\Delta \tau')_p = \frac{2G_0}{3kT} + \frac{2}{3}\ln(\dot{\varepsilon}/\dot{\varepsilon}_0). \tag{8-17}$$

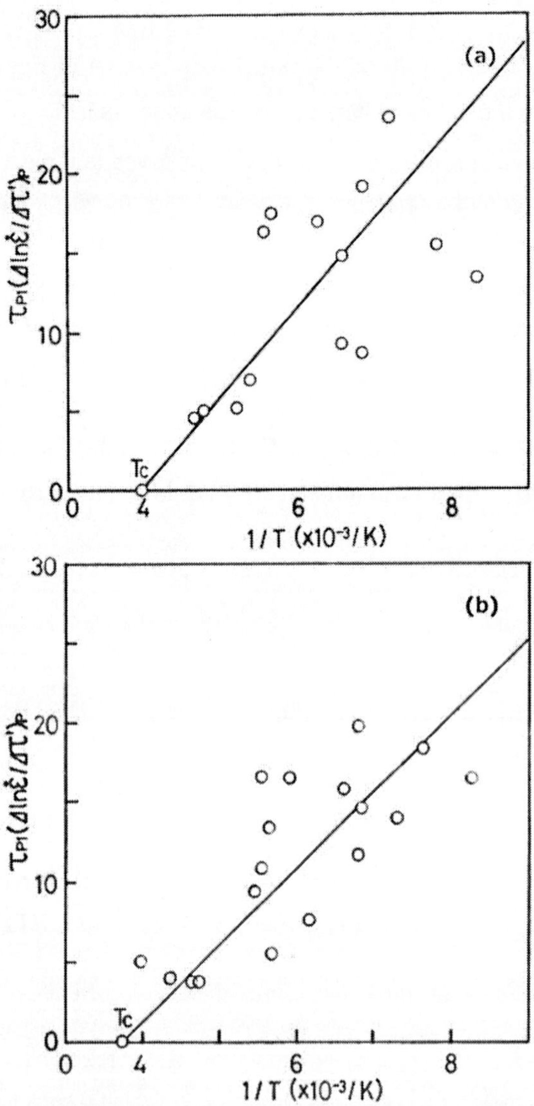

Figure 8-16. Linear plots of equation (8-17) for (a) KCl:Li$^+$ (0.50 mol.% in the melt) and (b) KCl:Na$^+$ (0.50 mol.% in the melt).

The $(\partial \ln \dot{\varepsilon}/\partial \tau^*)$ in equation (8-16) is used as a reciprocal of the $(\Delta\tau'/\Delta \ln \dot{\varepsilon})_p$ which is given by the difference between strain-rate sensitivities at the first plateau place and at the second one on the strain-rate sensitivity versus $\Delta\tau$ curve, as defined in figure 8-2. The $(\Delta\tau'/\Delta\ln\dot{\varepsilon})_p$ is assumed to be the strain-rate sensitivity due to impurities when a dislocation moves forward with the help of oscillation. The results of calculations for equation (8-17) are shown as the open circles in figure 8-16 (a) for KCl:Li^+ (0.50 mol.%) and (b) for KCl:Na^+ (0.50 mol.%). The slopes of straight lines in the figures are determined by the method of least squares. The temperature at which the line intersects the abscissa in the figure and $\tau_{p1}(\Delta\ln\dot{\varepsilon}/\Delta\tau')_p$ is zero, corresponds to the value of T_c. The Gibbs free energy for the breakaway of a dislocation from the impurity, which is obtained through the slope of straight line in figures 8-16 (a) and (b), is 0.73 eV for KCl:Li^+ and 0.62 eV for KCl:Na^+.

CONCLUSION

The following conclusions 1 and 2 were derived from the data analyzed in terms of the relative curve of strain-rate sensitivity of flow stress and the stress decrement due to oscillation.

1. The force-distance relation, which represents the interaction between a dislocation and an impurity, is investigated for KCl:Li^+ and KCl:Na^+ with respect to the three models (the SQ, the PA, and the TR) by the following three relationships: the relation between effective stress and temperature, the dependence of strain-rate sensitivity due to the impurities on temperature, and the proportionality of $\Delta H(T)$. As a result, the SQ seems to be the most suitable model of the three for KCl:Na^+ from the proportionality of $\Delta H(T)$. However, it is difficult to select the most suitable model of the three for KCl:Li^+ from the above-mentioned three relationships.
2. For the three force-distance relations, the values of T_c and $\Delta H(T_c)$ are almost the same, irrespective of the kind of force-distance relation for both the specimens. The value of T_c for KCl:Li^+ is obviously small in comparison with that for KCl:Na^+, whereas the value of $\Delta H(T_c)$ for KCl:Li^+ is nearly equal to that for KCl:Na^+. Furthermore, the Gibbs free energy for the breakaway of a dislocation from the monovalent impurity is 0.73 eV for KCl:Li^+ and 0.62 eV for KCl:Na^+. These are obtained through the slope of linear relationship between $1/T$ and $\tau_{p1}(\Delta\ln\dot{\varepsilon}/\Delta\tau')_p$.

   As for T_c, it tends to increase when the divalent cationic size is increasingly close to that of the K^+ from the small divalent cationic size side in KCl doped with various divalent cationic impurities (see sec. 4.3 of chap. 4 and sec. 5.3 of chap. 5). The same phenomenon as mentioned for KCl doped with the divalent additions may also be predicted for KCl doped with monovalent cationic ones.

REFERENCES

[1] Foreman, A. J. E.; Makin, M. J. *Phil. Mag.* 1966, *14*, 911.
[2] Suszyńska, M. *Kristall. Technik.* 1974, *9*, 1199.

[3] Dryden, J. S.; Morimoto, S.; Cook, J. S. *Phil. Mag.* 1965, *12*, 379.
[4] Stoloff, N. S.; Lezius, D. K.; Johnston, T. L. *J. Appl. Phys.* 1963, *34*, 3315.
[5] Chin, G. Y.; Van Uitert, L. G.; Green, M. L.; Zydzik, G. J.; Kometani, T. Y. *J. Amer. Ceram. Soc.* 1973, *56*, 369.
[6] Kataoka, T.; Yamada, T. *Jpn. J. Appl. Phys.* 1977, *16*, 1119.
[7] Guiu, F.; Langdon, T. G. *Phil. Mag.* 1974, *30*, 145.
[8] Johnston, W. G.; Gilman, J. J. *J. Appl. Phys.* 1959, *30*, 129.
[9] Ohgaku, T.; Takeuchi, N. *Phys. Status Solidi* (a) 1992, *134*, 397.
[10] Hikata, A.; Johnson, R. A.; Elbaum, C. *Phys. Rev. B* 1970, *2*, 4856.
[11] Ohgaku, T.; Takeuchi, N. *Phys. Status Solidi* (a) 1989, *111*, 165.
[12] Friedel, J. *Dislocations;* Pergamon Press; Oxford, 1964; p 224.
[13] Conrad, H. *Acta Metall.* 1966, *14*, 1631.
[14] Ono, K. *J. Appl. Phys.* 1968, *39*, 1803.
[15] Soullard, J.; Veyssiere, P. *Acta Metall.* 1983, *31*, 1177.
[16] Christian, J. W.; Masters, B. C. *Proc. R. Soc. A* 1964, *281*, 240.
[17] Conrad, H. *J. Metals* 1964, *16*, 582.
[18] Conrad, H. *J. Phys.* 1967, *45*, 581.
[19] Dey, B. N.; Tyson, W. R. *Phys. Status Solidi* (a) 1972, *9*, 215.
[20] Kataoka, T.; Uematsu, T.; Yamada, T. *Jpn. J. Appl. Phys.* 1978, *17*, 271.
[21] Granato, A. V.; Lücke, K. *J. Appl. Phys.* 1956, *27,* 583.
[22] Granato, A. V.; Lücke, K. *J. Appl. Phys.* 1956, *27,* 789.
[23] Ohgaku, T.; Takeuchi, N. *Phys. Status Solidi* (a) 1987, *102*, 293.
[24] Ohgaku, T.; Takeuchi, N. *Phys. Status Solidi* (a) 1988, *105*, 153.
[25] Sumino, K. *Jpn. Inst. Met.* 1972, *11*, 31 (in Japanese).

Chapter 9

Bending Angle at Which the Dislocation Breaks Away from the Obstacle in KCl Doped with Various Monovalent Impurities

9.1. Introduction

When a mobile dislocation is pressed against weak rigid obstacle by an effective stress (τ^*), the average spacing (L) of obstacles along the dislocation is approximately represented by [1]

$$L = \left(\frac{2 L_0{}^2 E'}{\tau^* b} \right)^{1/3}, \tag{9-1}$$

where L_0 is the average spacing of obstacles distributed randomly on the slip plane, is the line tension of the dislocations, and b is the magnitude of the Burgers vector. Equation (9-1) is termed the Friedel relation and has been applied to most weak obstacles to dislocation motion at low solute concentration. However, it has been often used without sufficient examination [2-4]. In this chapter, the applicable range of the Friedel relation is outlined on the basis of the $(L_0 / L)^2$ versus ϕ_0 curve, where ϕ_0 is the angle at which the dislocation embraces the impurity at a temperature of 0 K. And further, it is investigated whether the Friedel relation is appropriate for the interaction between a dislocation and the monovalent ions in various alkali halide crystals (NaCl:Br^-, NaBr: Cl^- or I^-, KCl: Li^+, Na^+, Br^- or I^-, and RbCl: Br^- or I^-), which is expected to cause the gradual hardening due to the cubic symmetry defects produced around the substitutional-type ions [5-7]. The applicability of it is examined on the basis of $(L_0 / L)^2$ versus the bending angle of dislocation by the dopant ion in each crystal.

9.2. THEORETICAL CONCEPT

Figure 9-1 shows the dislocation motion on the slip plane containing one type of weak obstacles (represented by open circles) such as impurities, where the mobile dislocation is pinned by a linear array of weak obstacles (A, B and C) and the dislocation segment (a solid line in figure 9-1) bows out between the obstacles. The dislocation breaks away from the obstacle B with the help of thermal activation. Then its bending angle at which the dislocation breaks away from the obstacle B is represented by ϕ. The dislocation passed obstacle B glides until it is pinned by next obstacle B' and reaches AB'C (a dashed line in figure 9-1). The advance of mobile dislocation is illustrated in figure 9-1.

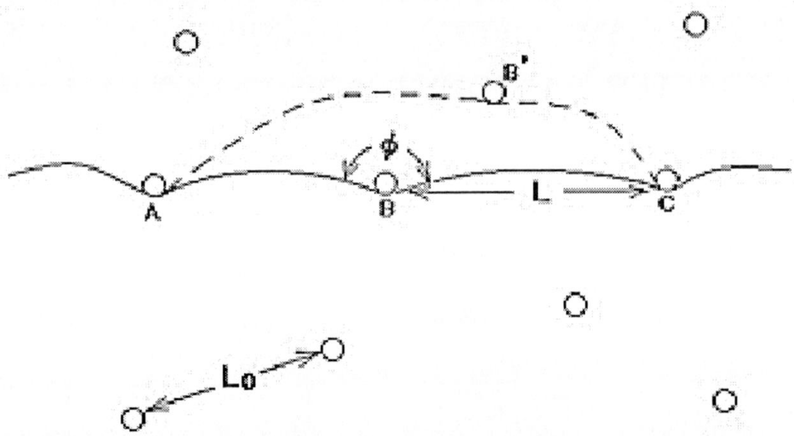

Figure 9-1. Dislocation held up at weak obstacles on the slip plane. Open circles represent one type of weak obstacles, such as impurities. The dislocation ABC (the solid line) moves forward to next AB'C (the dashed line) by the free flight motion.

The $(L_0/L)^2$ versus ϕ_0 (ϕ at the temperature of 0 K) curve for the Fleischer relation [6] is expressed as

$$\left(\frac{L_0}{L}\right)^2 = \frac{\pi - \phi_0}{2} \tag{9-2}$$

and that for the Friedel relation [1] is given by

$$\left(\frac{L_0}{L}\right)^2 = \cos\left(\frac{\phi_0}{2}\right). \tag{9-3}$$

The process of leading equations (9-2) and (9-3) is detailed in the section 5.1 of chapter 5.

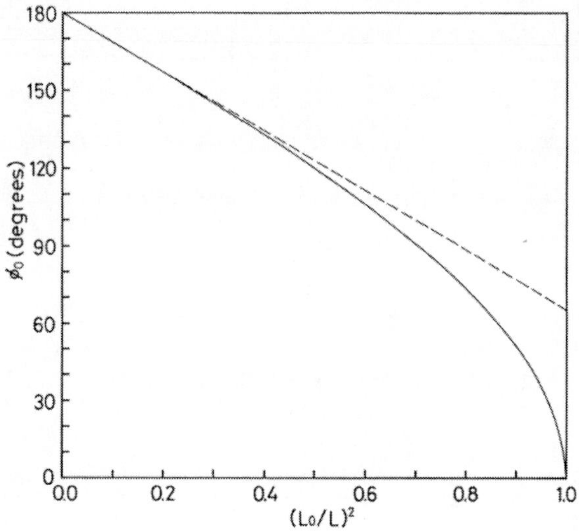

Figure 9-2. Relationship between $(L_0 / L)^2$ and ϕ_0. Solid line represents the curve obtained from the Friedel relation (see equation (9-3)); dashed line from the Fleischer relation (see equation (9-2)).

In figure 9-2, the relation between and for equation (9-2) and that for equation (9-3) are represented as a dashed and a solid lines, respectively. The Fleischer relation $(L_0^{\ 2} = L^2(\pi - \phi_0)/2)$ was suggested by Fleischer separately from the Friedel relation. The Friedel relation (equation (9-1)) was proposed originally by Friedel. Considering that dislocation AB'C successively breaks away also from obstacles A and C, the Fleischer relation is in essential agreement with the Friedel relation that the steady state is reached, i.e. a dislocation steadily overcomes thermal obstacles during steady and plastic deformation [8, 9]. The weak obstacles suitable for the Friedel relation will be simultaneously satisfied with the Fleischer relation, i.e. the applicable range of in equation (9-3) agrees with that of in equation (9-2). As above-mentioned result, it is considered that overlap of a solid and a dashed lines appear in figure 9-2. The solid line agrees with the dashed one in the range of $150° \leq \phi_0 \leq 180°$ and $0 \leq (L_0 / L)^2 \leq 0.28$. The result implies that the Friedel relation is suitable for the weak obstacle such as impedes the dislocation above of about 150 degrees during steady and plastic deformation.

9.3. Bending Angle

It is investigated whether the Friedel relation is appropriate for the interaction between a dislocation and the monovalent impurities in various alkali halide crystals with NaCl-type of structure. This is examined from the bending angle ϕ_0. The values of under the effective shear stress is given by (see sec. 5.4 of chap. 5)

$$\cos\frac{\phi_0}{2} = \left(\frac{\tau^* L_0 b}{2E}\right)^{2/3}. \tag{9-4}$$

Equation (9-1) will be transformed to

$$L_0 = \left(\frac{\tau^* L^3 b}{2E}\right)^{1/2}. \tag{9-5}$$

Substituting of equation (9-5) in equation (9-4) gives

$$\cos\frac{\phi_0}{2} = \frac{\tau^* Lb}{2E}, \tag{9-6}$$

that is to say,

$$\frac{\phi_0}{2} = \cos^{-1}\left(\frac{\Delta G_0}{2\mu b^3}\right), \tag{9-7}$$

where is the activation energy for overcoming of an impurity by a dislocation and can be represented by $\tau^* Lb^2$ and is the shear modulus. The line tension is replaced by μb^2. The reciprocal of is given by [10]

$$\frac{1}{\mu} = S_{44} + 4\left(S_{11} - S_{12} - \frac{1}{2}S_{44}\right)\left(l_1^2 l_2^2 + l_2^2 l_3^2 + l_3^2 l_1^2\right), \tag{9-8}$$

where S_{11}, S_{12} and S_{44} are the adiabatic elastic compliances and l_1, and are the direction cosine of stress-axis. The values $l_1 = 1/\sqrt{2}$, and are used for the [110] direction. The adiabatic elastic compliances S_{11}, S_{12} and S_{44} for cubic crystals are related to the elastic stiffnesses C_{11} , and by the following equations [11]:

$$S_{11} = \frac{C_{11} + C_{12}}{(C_{11} + 2C_{12})(C_{11} - C_{12})}, \tag{9-9}$$

$$S_{12} = \frac{-C_{12}}{(C_{11} + 2C_{12})(C_{11} - C_{12})}, \tag{9-10}$$

$$S_{44} = \frac{1}{C_{44}}. \tag{9-11}$$

Data used in calculations of is given in tables 9-1 and 9-2 for the several kinds of alkali halide crystals. By using the values of derived from equations (9-8) to (9-11) and in table 9-3, it is possible to obtain the bending angle from equation (9-7), whereupon the magnitude of the Burgers vector of a dislocation with $\boldsymbol{b}=r$ [110] is (i.e., $b=\sqrt{2}r$ whereof is the interionic distance at 0 K [12]). As shown in table 9-3, these calculations for all crystals are within $173° \le \phi_0 < 180°$, where the two curves (a solid and a dashed lines) in figure 9-2 overlap each other.

Table 9-1. Values of the elastic stiffnesses C_{ij} ($\times 10^{10}$ Pa) for each matrix of various crystals. The data are obtained at the temperature T (K) indicated in the column

Single crystal	C_{11}	C_{12}	C_{44}	T
NaBr [13]	4.800	0.986	1.070	4.2
RbCl [13]	4.297	0.649	0.493	4.2

Table 9-2. Values of the elastic compliances S_{ij} ($\times 10^{-11}$ Pa^{-1}) for each matrix of various crystals[a]. The data are obtained at the temperature T (K) indicated in the column

Single crystal	S_{11}	S_{12}	S_{44}	T
NaCl [14]	2.0	−0.31	7.4	0
NaBr	2.24	−0.382	9.35	4.2
KCl [14]	2.05	−0.32	15.0	0
RbCl	2.42	−0.318	20.28	4.2

[a] The data for NaBr and RbCl in table 9-2 is derived from equations (9-9) to (9-11).

Table 9-3. Bending angle ϕ_0 (degrees) and $(L_0/L)^2$ derived from equations (9-3), (9-7) and input parameters[b] (the shear modulus ($\times 10^{10}$ Pa) for [110] direction, the magnitude of the Burgers vector b (Å) ($=\sqrt{2}r$) at 0 K and the Gibbs free energy (eV) for the dislocation motion) for the NaCl-type alkali halides

Single crystal	ϕ_0	$(L_0/L)^2$	(L/L_0)	μ	b	ΔG_0
$NaCl:Br^-$(0.1, 0.5, 1 mol%)	176.7	0.029	5.87	1.66	3.944	0.37 [15]
$NaBr:Cl^-$ (1.0 mol%)	176.3	0.032	5.59	1.371	4.178	0.40 [16]
$NaBr:I^-$ (1.0 mol%)	175.7	0.038	5.15	1.371	4.178	0.47 [16]
$KCl:Li^+$ (0.5 mol%)	177.4	0.023	6.59	1.01	4.407	0.73[c]
$KCl:Na^+$ (0.5 mol%)	177.7	0.020	7.07	1.01	4.407	0.62[c]
$KCl:Br^-$ (1, 2 mol%)	174.7~ 175.8	0.037~ 0.046	4.65~ 5.19	1.01	4.407	0.4~0.5 [17]
$KCl:I^-$ (0.2 mol%)	174.7~ 175.8	0.037~ 0.046	4.65~ 5.19	1.01	4.407	0.4~0.5 [17]
$RbCl:Br^-$ (1.0 mol%)	174.0	0.053	4.36	0.776	4.609	0.50 [18]
$RbCl:I^-$ (1.0 mol%)	173.0	0.061	4.05	0.776	4.609	0.58 [18]

[b] The values of shear modulus in table 9-3 are approximately estimated as each matrix of various crystals at about 0 K.

[c] The values of ΔG_0 for KCl: Li^+ or Na^+ are obtained in the section 8.8 of chapter 8.

This means that the Friedel relation is applicable to the interaction between a dislocation and the monovalent impurities in the crystals. Then, the values of are derived from equation (9-3) and can be found within $0.020 \le (L_0/L)^2 \le 0.061$, namely, $4.05 \le L/L_0 \le 7.07$ in table 9-3. The value of becomes lower at lower as given in table 9-3. Bending angle is small, when the dopant ion is strong obstacle to dislocation motion (I^- ions in RbCl are the strongest of all crystals here). This leads to the result that L approaches to the length of L_0 by the curvature of dislocation. Value of (L/L_0), therefore, is characteristic for each crystal, although the parameters L and L_0 depend on doping concentration.

CONCLUSION

On the basis of the versus ϕ_0 curves of the Friedel and the Fleischer relations, it was found out that the Friedel relation between the effective stress and the average length of dislocation segments, as well as the Fleischer relation ($L_0 = L^2(\pi - \phi_0)/2$), is suitable for the interaction between a dislocation and the monovalent impurities in various alkali-halide single crystals: NaCl:Br^-, NaBr: Cl^- or I^-, KCl: Li^+, Na^+, Br^- or I^- and RbCl: Br^- or I^-. This is because the dopant ions are the weak obstacles such as impede the mobile dislocation at bending angle above about 150 degrees, where the Friedel relation agrees with the Fleischer relation. Furthermore, the values of (L/L_0) were found to be within 4.05 to 7.07 for the crystals.

REFERENCES

[1] Friedel, J. *Dislocations;* Pergamon Press; Oxford, 1964; p 224.
[2] Soullard, J.; Veyssiere, P. *Acta Metall.* 1983, *31*, 1177.
[3] Ohgaku, T.; Takeuchi, N. *Phys. Status Solidi* (a) 1992, *134*, 397.
[4] Ohgaku, T.; Matsunaga, T. *Pro. Int. Meeting "Dislocations 2008"* in IOP Conf. Series: *Mat. Sci. Eng.* 2009, *3*, 012021.
[5] Fleischer, R. L. *Acta Metall.* 1962, *10*, 835.
[6] Fleischer, R. L.; Hibbard Jr., W. R. *The Relation between Structure and Mechanical Properties of Metals, National Physical Laboratory*, Her Majesty's Stationary Office; London, 1963; Vol. 1, pp 261-297.
[7] Johnston, W. G.; Nadeau, J. S.; Fleischer, R. L. *J. Phys. Soc. Japan Suppl. I* 1963, *18*, 7.
[8] Sumino, K. *Jpn. Inst. Met.* 1972, *11*, 31 (in Japanese).
[9] Sumino, K. *Plasticity of Crystals,* Maruzen; Tokyo, 1977; pp 365-374 (in Japanese).
[10] Igata, N. *Strength of Materials,* Baifukan; Tokyo, 1991; pp 8-12 (in Japanese).
[11] Sirdeshmukh, D. B.; Sirdeshmukh, L.; Subhadra, K. G. *Alkali Halide,* (edited by Hull, R.; Osgood, Jr., R. M.; Sakaki, H.; Zunger, A.) Springer-Verlag; Berlin Heidelberg, 2001; p 17.

[12] Sirdeshmukh, D. B.; Sirdeshmukh, L.; Subhadra, K. G. *Alkali Halide,* (edited by Hull, R.; Osgood, Jr., R. M.; Sakaki, H.; Zunger, A.) Springer-Verlag; Berlin Heidelberg, 2001; p 7.

[13] Varshni, Y. P.; Konti, A. *Phys. Rev. B* 1972, *6*, 1532.

[14] Hart, S. *Brit. J. Appl. Phys.* (*J. Phys. D*) ser. 2, 1968; *1*, 1285.

[15] Ohgaku, T.; Teraji, H. *Phys. Status Solidi* (a) 2001, *187*, 407.

[16] Ohgaku, T.; Sakuda, M. *Mater. Sci. Engng. A* 2003, *357*, 55.

[17] Ohgaku, T.; Takeuchi, N. *Meeting abstracts of the Physical Society of Japan* Vol. 1991, No. 2 (199110311) p 54.

[18] Ohgaku, T.; Hashimoto, K. *Mater. Sci. Engng. A* 2005, *400*, 401.

Chapter 10

INFLUENCE OF A STRAIN ON THE INTERACTION BETWEEN A DISLOCATION AND THE IMPURITIES

10.1. INTRODUCTION

The stress required to overcome dislocations threading the glide plane by a dislocation has been regarded as the flow stress, which is expressed as a function of the density of forest dislocations, for pure crystals. This is generally supported at present [1]. However, work-hardening theory, by which the dependence of flow stress and strain on dislocation density is satisfactorily completed from both the sides of quantity and of mechanism, is not established yet [1, 2].

In this chapter, the strain-rate cycling tests associated with the oscillation are carried out for KCl:Li^+ (0.50 mol.% in the melt) or Na^+ (0.50 mol.% in the melt) single crystals in stage I (easy glide region), stage II (linear hardening region), and stage III (dynamic recovery region) of stress-strain curve. On the basis of the variation of the strain-rate sensitivity versus stress decrement curve with shear strain, it is discussed whether dislocation multiplication is influenced by Li^+ and Na^+ ions in the specimens. The effective stress due to impurities and T_c are further examined in the different plastic deformation regions. The purpose is to investigate the influence of shear strain on the interaction between a dislocation and impurities. The analysis is discussed according to the variation of the relative curve of strain-rate sensitivity and stress decrement with shear strain.

10.2. INFLUENCE OF STRAIN ON THE STRAIN-RATE SENSITIVITY VERSUS STRESS DECREMENT CURVE

10.2.1. Relation between the Strain-Rate Sensitivity and the Stress Decrement

Figure 10-1 shows the relation between strain-rate sensitivity and stress decrement due to the oscillation at various shear strains for KCl:Na^+ (0.50 mol.% in the melt) at 180 K. The strain-rate sensitivity varies with stress decrement at a given shear strain and the relative curve of strain-rate sensitivity and stress decrement shifts upward with increasing shear strain.

The same phenomenon as figure 10-1 is also observed for KCl:Li^{+} (0.50 mol.% in the melt). The first bending point at low stress decrement is denoted by τ_{p1} and the second one at high stress decrement by τ_{p2}. τ_{p1} is considered to represent the effective stress due to only one type of the impurities which lie on the dislocation when the dislocation moves forward with the help of oscillation (see sec. 2.7 of chap. 2). At the second plateau region more than τ_{p2} on the curve, the obstacles to the dislocation motion are considered to be only forest dislocations and the impurities no longer act as obstacles. Therefore, the above-mentioned phenomenon that the relative curve of strain-rate sensitivity and stress decrement shifts upward with increasing shear strain may be caused by the part of the strain-rate sensitivity due to dislocation cutting process as described for KCl:Sr^{2+} (see sec. 2.4 of chap. 2). Because, when the waiting time for the thermal activation equals at two kinds of obstacles such as impurities and forest dislocations, the reciprocal of the activation volume, υ^{*}, is given by [3, 4]

$$\frac{1}{\upsilon^{*}} = \frac{1}{\upsilon_p^{*}} + \frac{1}{\upsilon_s^{*}}, \tag{10-1}$$

where the suffixes of p and s are due to impurities and dislocation cutting process, respectively. Here the activation volume is given by

$$\upsilon^{*}(= bLd) = kT\left(\frac{\Delta \ln \dot{\varepsilon}}{\Delta \tau'}\right)_T, \tag{10-2}$$

where d is an activation distance.

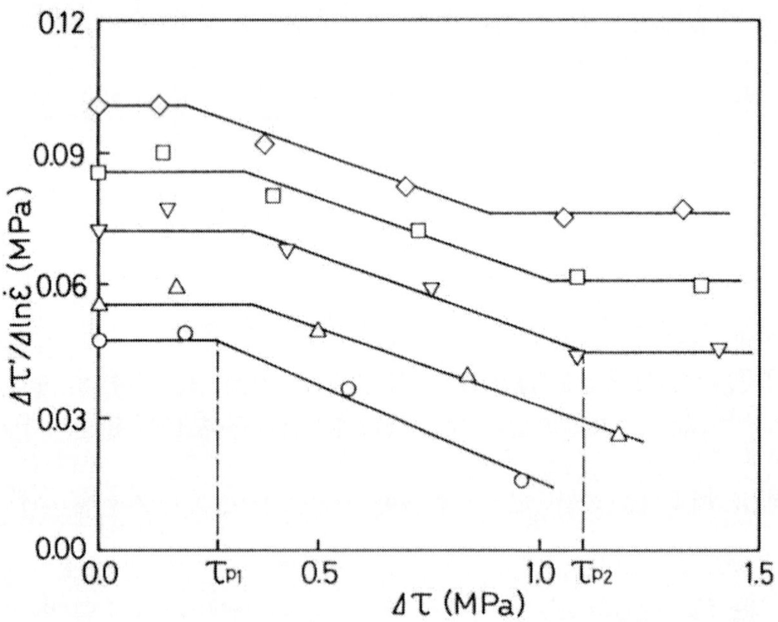

Figure 10-1. Relation between the strain-rate sensitivity and the stress decrement for KCl:Na^{+} (0.50 mol.% in the melt) at 180 K and strains: ε = (○) 6 %, (Δ) 12 %, (▽) 17 %, (□) 20 %, (◇) 23 %.

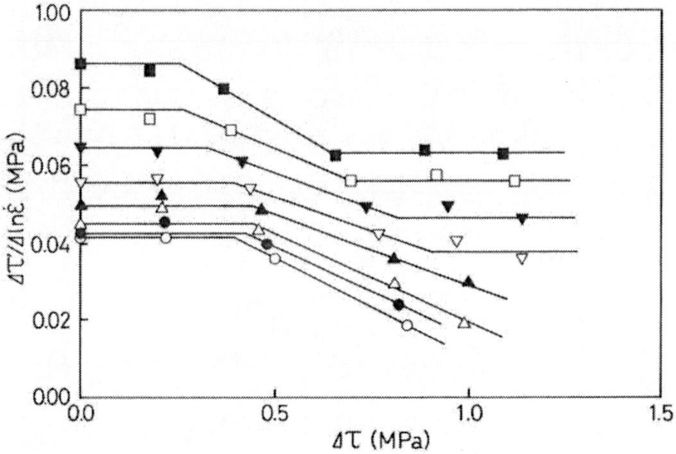

Figure 10-2. Relation between the strain-rate sensitivity and the stress decrement for KCl:Na$^+$ (0.50 mol.% in the melt) at 169 K and strains: ε = (○) 6 %, (●) 8 %, (Δ) 10 %, (▲) 12 %, (▽) 14 %, (▼) 16 %, (□) 18 %, (■) 20 %.

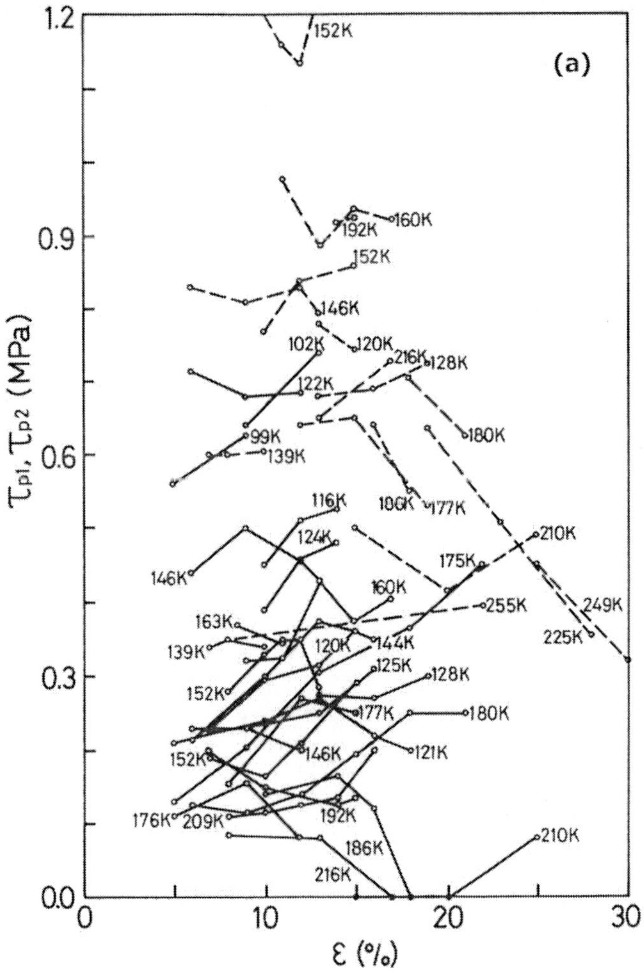

Figure 10-3. (Continued).

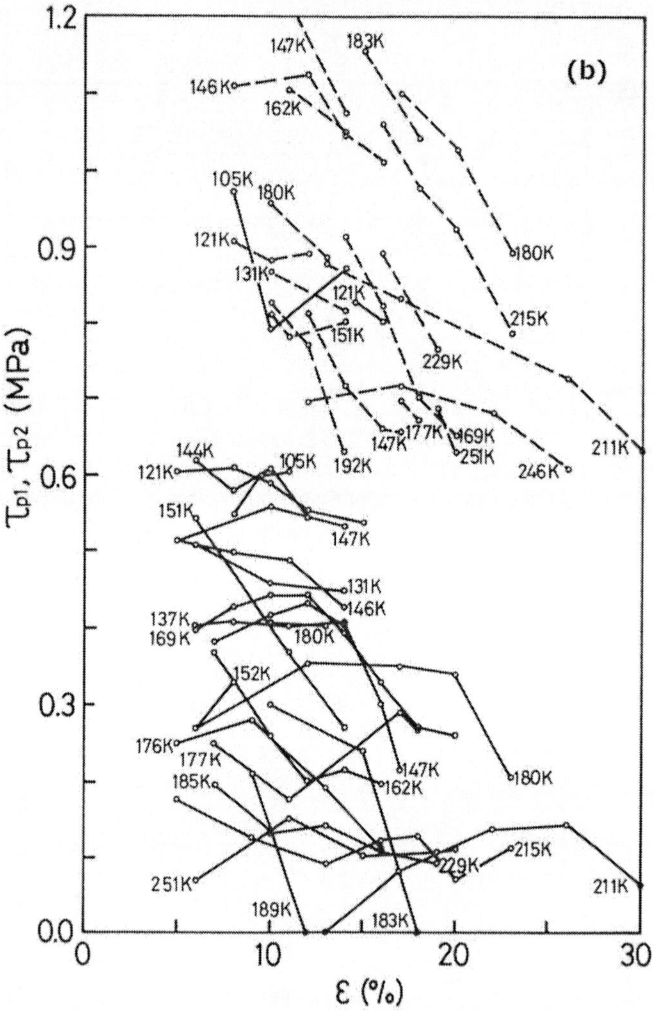

Figure 10-3. Dependence of (—) τ_{p1} and (- - -) τ_{p2} on the shear strain at various temperatures for (a) $KCl:Li^+$ (0.50 mol.% in the melt) and (b) $KCl:Na^+$ (0.50 mol.% in the melt).

10.2.2. Dependence of τ_{P1} and τ_{P2} on the Shear Strain

The relation between strain-rate sensitivity and stress decrement at various shear strains is shown in figure 10-2 for $KCl:Na^+$ at 169 K. Both the τ_{p1} and the τ_{p2} almost decrease with increasing shear strain. The same phenomenon as figure 10-2 also appears on the strain-rate sensitivity versus stress decrement curves for $KCl:I^-$ (0.5 mol.% in the melt) at 205 K [5] and $KCl:Sr^{2+}$ (0.05 mol.% in the melt) at 200 K (see figure 2-10 in sec. 2.4 of chap. 2). The dependence of τ_{p1} and τ_{p2} on shear strain at various temperatures is shown in figure 10-3 (a) for $KCl:Li^+$ and figure 10-3 (b) for $KCl:Na^+$. The τ_{p1} and the τ_{p2} decrease with increasing shear strain at almost all the temperatures for $KCl:Na^+$. However, it cannot be discerned for $KCl:Li^+$.

10.3. Critical Temperature and the Effective Stress

In the section 8.7 of chapter 8, the square force-distance relation between a dislocation and the impurity seems to be the most suitable of the three [6], which are a square one, a parabolic one and a triangular one, taking account of the Friedel relation [7] for $KCl:Na^+$. Then, the relation formula of τ_{p1} and temperature, which reveals the force-distance relation between a dislocation and an impurity (see sec. 2.5 of chap. 2), is given by

$$\left(\frac{\tau_{p1}}{\tau_{p0}}\right)^{2/3} = 1 - \left(\frac{T}{T_c}\right), \tag{10-3}$$

where τ_{p0} is the effective stress due to the impurities without thermal activation and T_c is the critical temperature at which τ_{p1} is zero. The values of τ_{p0} and T_c can be obtained through the research on the linear plots of equation (10-3). τ_{p0} is determined by extrapolating the linear relationship of τ_{p1} and temperature to 0 K. Since the τ_{p1} for $KCl:Na^+$ decreases with increasing shear strain at almost all the temperatures in figure 10-3 (b), one investigates the values of T_c and τ_{p0} for $KCl:Na^+$ on the basis of the relation between τ_{p1} and temperature in the different plastic deformation region. The τ_{p1} in stage I is obtained from the strain-rate sensitivity versus stress decrement curve at the smallest shear strain in stage I and the τ_{p1} in stage II from the curve at the largest shear strain in stage II. The results for T_c and τ_{p0} are given in table 10-1. The T_c in stage I is larger than that in stage II and the τ_{p0} in stage I is slightly larger than that in stage II.

Table 10-1. Values of T_c and τ_{p0} in the different plastic deformation region for $KCl:Na^+$ (0.50 mol.% in the melt)

Plastic deformation region	T_c (K)	τ_{p0} (MPa)
Stage I	269	1.36
Stage II	260	1.34

10.4. Strain-Rate Sensitivity at the Second Plateau Place on the Strain-Rate Sensitivity Versus Stress Decrement Curve

Three-stage strain hardening has been clearly established as the characteristic behavior of rock salt structure single crystals deforming predominantly by single glide [8]. The three-stage strain hardening is also obtained for KCl [9, 10]. In this section, it is observed above 124 K for $KCl:Li^+$ and above 105 K for $KCl:Na^+$. The strain hardening curves are shown in figure 10-4 (a) for $KCl:Li^+$ at 160 K, (b) for $KCl:Na^+$ at 169 K and (c) for KCl at 142 K. With respect to the strain-rate sensitivity versus stress decrement curve, the strain-rate sensitivity at the second plateau place on the curve is investigated at the following two shear strains. One is the boundary between stage I and II, namely, A in figure 10-4. The other is that between stage

II and III, namely, B in figure 10-4. The former strain-rate sensitivity is hereafter termed $(\Delta\tau'/\Delta\ln\dot{\varepsilon})_{I\text{-}II}$ and the latter one $(\Delta\tau'/\Delta\ln\dot{\varepsilon})_{II\text{-}III}$.

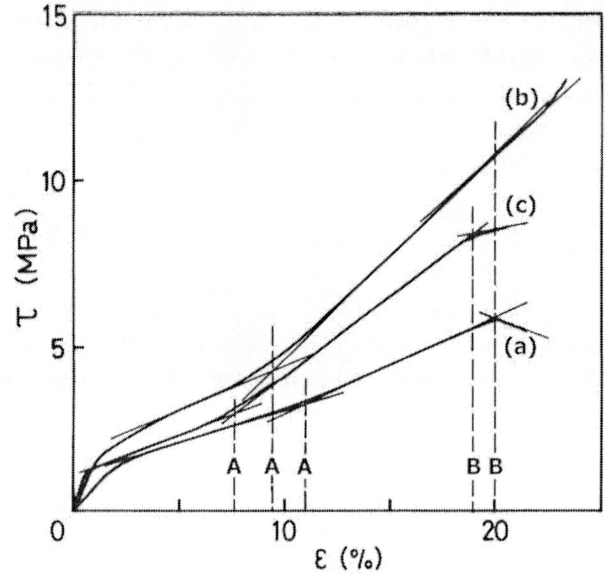

Figure 10-4. Shear stress-shear strain curves for (a) KCl:Li^+ (0.50 mol.% in the melt) at 160 K, (b) KCl:Na^+ (0.50 mol.% in the melt) at 169 K and (c) KCl at 142 K.

The result for the strain-rate sensitivity is shown in figure 10-5 (a) for KCl:Li^+ and KCl and (b) for KCl:Na^+ and KCl. The open symbols correspond to that for KCl:Li^+ and KCl:Na^+ and the solid ones that for KCl. The curves in figure 10-5 (a) and (b) are to guide the reader's eye. Comparing the $(\Delta\tau'/\Delta\ln\dot{\varepsilon})_{I\text{-}II}$ and the $(\Delta\tau'/\Delta\ln\dot{\varepsilon})_{II\text{-}III}$ for KCl:Li^+ and KCl:Na^+ with those for KCl, great difference is not seen between KCl:Li^+ and KCl, whereas those for KCl:Na^+ are obviously larger as against those for KCl. That is, the Li^+ ions in KCl:Li^+ have no influence on either $(\Delta\tau'/\Delta\ln\dot{\varepsilon})_{I\text{-}II}$ and $(\Delta\tau'/\Delta\ln\dot{\varepsilon})_{II\text{-}III}$. However, the Na^+ ions in KCl:Na^+ seem to make those large. Therefore, the forest dislocation density seems to increase by the Na^+ ions in KCl:Na^+ within the temperature range. But the Li^+ ions in KCl:Li^+ do not influence it. Furthermore, the variation of strain-rate sensitivity (see the marked part with grey circle in figure 10-6) at the second plateau place on the strain-rate sensitivity versus stress decrement curve with shear strain (i.e. $\Delta(\Delta\tau'/\Delta\ln\dot{\varepsilon})/\Delta\varepsilon$), where the obstacles to the dislocation motion are only forest dislocations and the impurities no longer act as obstacles (see sec. 8.5 of chap.8), is investigated for the three kinds of specimens in figure 10-7 (a) and (b). The variation of strain-rate sensitivity at second plateau place with shear strain ($\Delta(\Delta\tau'/\Delta\ln\dot{\varepsilon})/\Delta\varepsilon$) is considered to be the increase of forest dislocation density with the shear strain. The relations of temperature and $\Delta(\Delta\tau'/\Delta\ln\dot{\varepsilon})/\Delta\varepsilon$ in stage I, II and III are represented by a circle, a triangle and a square, respectively. The $\Delta(\Delta\tau'/\Delta\ln\dot{\varepsilon})/\Delta\varepsilon$ in stage II is obviously larger than that in stage I at a given temperature. This phenomenon may suggest that there is only a slight increase in forest dislocation density with the shear strain in stage I but there is a rapid increase with that in stage II. The $\Delta(\Delta\tau'/\Delta\ln\dot{\varepsilon})/\Delta\varepsilon$ in stage III is larger than that in stage II at a given temperature. The $\Delta(\Delta\tau'/\Delta\ln\dot{\varepsilon})/\Delta\varepsilon$ in stage I and II increases with decreasing temperature for the three kinds of specimens. Both $(\Delta\tau'/\Delta\ln\dot{\varepsilon})_{I\text{-}II}$ and $(\Delta\tau'/\Delta\ln\dot{\varepsilon})_{II\text{-}III}$ also increase with decreasing temperature for $KCl:Li^+$ and $KCl:Na^+$. Thus, when the specimens are deformed by the compression test in stage I and II within the temperature, it will be predicted that the forest dislocation density at a given shear strain increases with decreasing temperature. When Ge single crystals were deformed by tensile test at a constant strain rate of 1.1×10^{-4} s^{-1} at 873 to 993 K, the dislocation density on the surface of specimens increases with decreasing temperature in the same deformation region of stage I and II [11]. The experimental result for Ge single crystals agrees with the above-mentioned prediction.

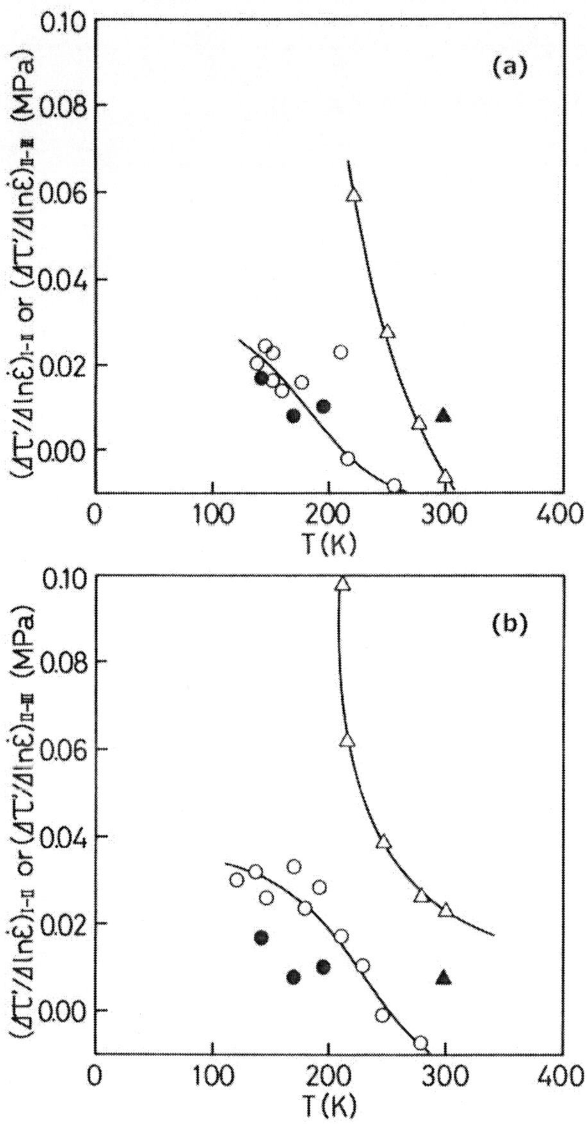

Figure 10-5. Dependence of $(\Delta\tau'/\Delta\ln\dot{\varepsilon})_{I\text{-}II}$ and $(\Delta\tau'/\Delta\ln\dot{\varepsilon})_{II\text{-}III}$ on the temperature for (a) $KCl:Li^+$ (0.50 mol.% in the melt) and KCl and (b) $KCl:Na^+$ (0.50 mol.% in the melt) and KCl. $(\Delta\tau'/\Delta\ln\dot{\varepsilon})_{I\text{-}II}$: (○) for $KCl:Li^+$ and $KCl:Na^+$ and (●) for KCl. $(\Delta\tau'/\Delta\ln\dot{\varepsilon})_{II\text{-}III}$: (Δ) for $KCl:Li^+$ and $KCl:Na^+$ and (▲) for KCl.

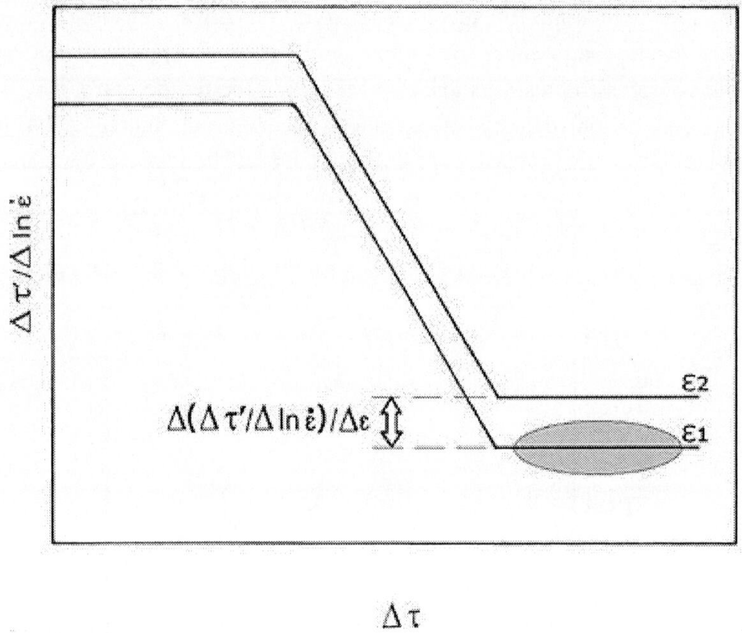

Figure 10-6. Illustration of relationship between the strain-rate sensitivity of flow stress and the stress decrement at a given strain, ε . ($\varepsilon_1 < \varepsilon_2$).

According to figure 10-7 (a) and (b), $\Delta(\Delta\tau'/\Delta\ln\dot{\varepsilon})/\Delta\varepsilon$ depends greatly on the temperature in stage II in contrast to that in stage I for the three kinds of specimens. It can be also observed that the value of $\Delta(\Delta\tau'/\Delta\ln\dot{\varepsilon})/\Delta\varepsilon$ for KCl:Li^+ is almost the same as that for KCl but the value of $\Delta(\Delta\tau'/\Delta\ln\dot{\varepsilon})/\Delta\varepsilon$ for KCl:Na^+ is remarkably larger as compared with that for KCl in stage I, II and III. Consequently, it seems that the Na^+ ions in KCl:Na^+ contribute to dislocation multiplication in the three stages within the temperature, whereas the Li^+ ions in KCl:Li^+ do not contribute to it.

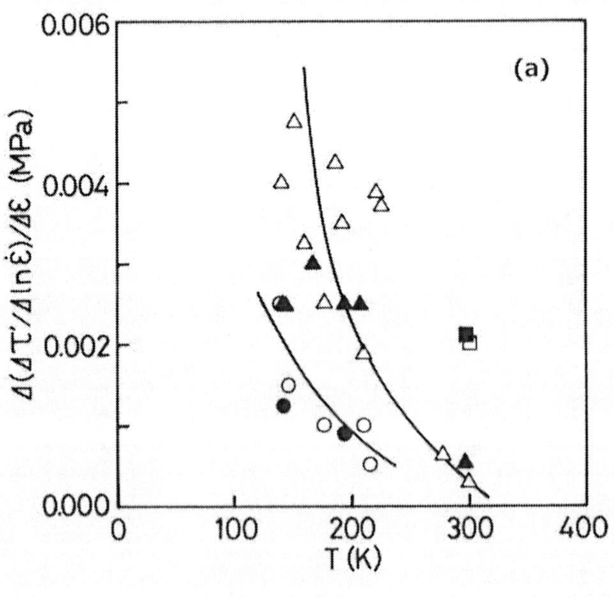

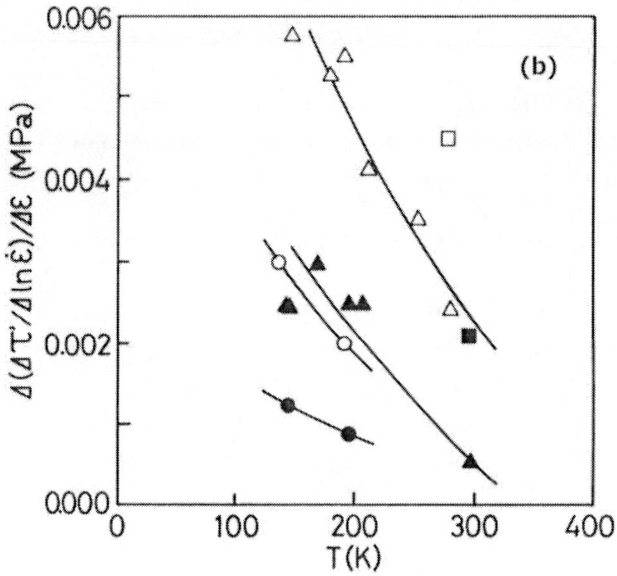

Figure 10-7. Dependence of $\Delta(\Delta\tau'/\Delta\ln\dot{\varepsilon})/\Delta\varepsilon$ on the temperature for (a) KCl:Li^+ (0.50 mol.% in the melt) and KCl and (b) KCl:Na^+ (0.50 mol.% in the melt) and KCl in the different plastic deformation region: (○) for KCl:Li^+ and KCl:Na^+ and (●) for KCl in stage I; (Δ) for KCl:Li^+ and KCl:Na^+ and (▲) for KCl in stage II; (□) for KCl:Li^+ and KCl:Na^+ and (■) for KCl in stage III.

Conclusion

On the basis of the strain-rate sensitivity versus stress decrement curve, influence of strain on the interaction between a dislocation and the impurities was analyzed as the following conclusions 1 and 2.

1. The phenomenon that both τ_{p1} and τ_{p2} decrease with increasing shear strain is observed at almost all the temperatures for KCl:Na^+. However, it cannot be discerned for KCl:Li^+. As a result, the T_c in stage I is larger than that in stage II and the τ_{p0} in stage I is slightly larger than that in stage II for KCl:Na^+ from the dependence of effective stress on temperature.
2. Comparing the $(\Delta\tau'/\Delta\ln\dot{\varepsilon})_{I\text{-}II}$ and $(\Delta\tau'/\Delta\ln\dot{\varepsilon})_{II\text{-}III}$ for KCl:Li^+ with those for KCl, great difference is not seen. The value of $\Delta(\Delta\tau'/\Delta\ln\dot{\varepsilon})/\Delta\varepsilon$ for KCl:Li^+ is also almost the same as that for KCl. However, both the $(\Delta\tau'/\Delta\ln\dot{\varepsilon})_{I\text{-}II}$ and the $(\Delta\tau'/\Delta\ln\dot{\varepsilon})_{II\text{-}III}$ for KCl:Na^+ are obviously larger as against those for KCl within the temperature. The value of $\Delta(\Delta\tau'/\Delta\ln\dot{\varepsilon})/\Delta\varepsilon$ for KCl:Na^+ is also larger than that for KCl in stage I, II and III of stress-strain curve. Therefore, it seems that the Na^+ ions in KCl:Na^+ make the forest dislocation density increased and contribute to dislocation multiplication in the three stages within the temperature. But the Li^+ ions in KCl:Li^+ do not influence the forest dislocation density.

REFERENCES

[1] Marukawa, K. *Materia Japan* 2001, *40*, 655, (in Japanese).

[2] Sumino, K. In *Strength of Crystals*, (edited by Hasiguti, R. R.; Chikazumi, S.) Asakurashoten; Tokyo, 1968; p 129 (in Japanese).

[3] Kocks, U. F. *Trans. JIM Suppl.* 1968, *9*, 1.

[4] Kocks, U. F.; Argon, A. S.; Ashby, M. F. *Thermodynamics and Kinetics of Slip*, (edited by Chalmers, B.; Christian, J. W.; Massalski, T. B.) Pergamon Press; New York, 1975; Progress in Materials Science Vol. 19, p 160.

[5] Ohgaku, T.; Takeuchi, N. *Phys. Status Solidi* (a) 1992, *134*, 397.

[6] Foreman, A. J. E.; Makin, M. J. *Phil. Mag.* 1966, *14*, 911.

[7] Friedel, J. *Dislocations;* Pergamon Press; Oxford, 1964; p 224.

[8] Sprackling, M. T. *The Plastic Deformation of Simple Ionic Crystals*, (edited by Alper, A. M.; Margrave, J. L.; Nowick, A. S.) Academic Press; London, 1976; p 203.

[9] Alden, T. H. *Trans. Met. Soc. AIME* 1964, *230*, 649.

[10] Davis, L. A.; Gordon, R. B. *J. Appl. Phys.* 1969, *40*, 4507.

[11] Kojima, K.; Sumino, K. *Cryst. Latt. Defects* 1971, *2*, 147.

Chapter 11

INFLUENCE OF THE DIVALENT IMPURITIES ON DISLOCATION DENSITY OF KCL DOPED WITH VARIOUS DIVALENT IMPURITIES

11.1. INTRODUCTION

The measurement of dislocation density in a crystal has been carried out by the etch pit technique. However, it is difficult to resolve the individual etch pits for high dislocation density. X-ray topography has also been used, but there is lack of resolution in the photograph for large deformation. Therefore, the specimens for these methods are limited to low dislocation density. Although the electron microscopy can be used to detect dislocations, it provides the information for a thin specimen but not for bulk. The measurement of strain-rate sensitivity under application of ultrasonic oscillatory stress gives the information for bulk during plastic deformation, where the interaction between a mobile dislocation and impurities, including the change in forest dislocation density due to the impurities with shear strain, has been obtaining from the relative position of strain-rate sensitivity versus stress decrement curves due to oscillation until this chapter 11. From the measurement, relative curve of strain-rate sensitivity and stress decrement has been obtained as given in figure 11-1. Figure 11-1 shows the curve for KCl:Ba^{2+} (0.050 mol.% in the melt). The curve for KCl doped with the divalent impurities has two bending points and two plateau regions. The relative positions of the curves reflect the influence of ultrasonic oscillation on the dislocation motion on the slip plane containing many impurities and a few forest dislocations. It is proposed that the variation of strain-rate sensitivity at the second plateau place on the curve with shear strain, $\Delta(\Delta\tau'/\Delta\ln\dot{\varepsilon})/\Delta\varepsilon$, indicates a change in forest dislocation density with shear strain (see sec. 10.4 of chap. 10). In this chapter, single crystals of nominally pure KCl and KCl doped with Mg^{2+}, Ca^{2+}, Sr^{2+} or Ba^{2+} were deformed by the same method as described in chapter 2 at 77 to 254 K; during the compression tests strain-rate cycling was conducted in association with ultrasonic oscillation. The purpose of it is to investigate in more detail whether the forest dislocation density is influenced by the concentration and the size of divalent impurities in KCl during plastic deformation. It is also considered in more detail whether the mobile dislocation density is influenced by divalent impurities in KCl during plastic deformation, which is analyzed in terms of the relative curve of strain-rate sensitivity and stress decrement due to oscillation. This means seems useful until the dislocation density is below about 10^{10} cm^{-2} (see sec. 4.4 of chap. 4).

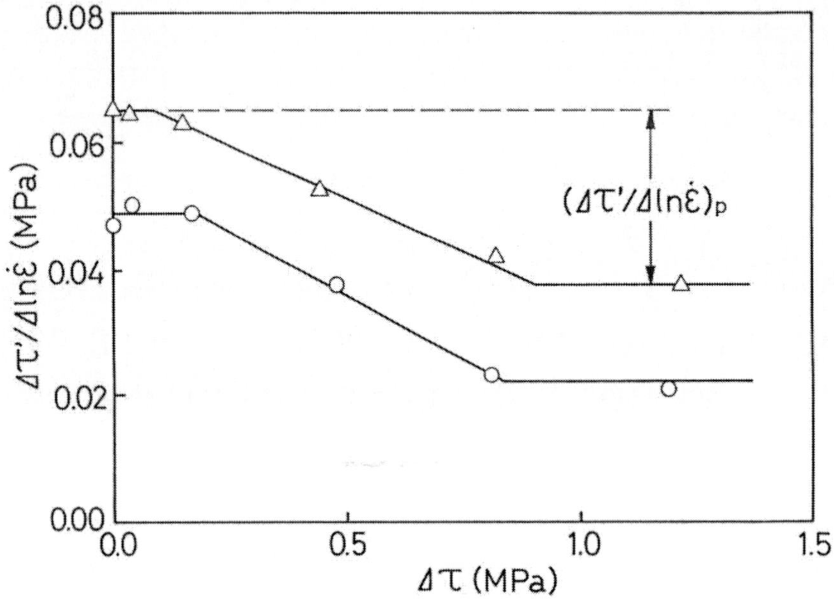

Figure 11-1. Strain-rate sensitivity versus stress decrement curve for KCl:Ba^{2+} (0.050 mol.%) at 178 K. ε : (○) 10 %, (Δ) 13 %. $(\Delta\tau'/\Delta\ln\dot{\varepsilon})_p$ is assumed to be the strain-rate sensitivity due to impurities.

11.2. Influence of the Divalent Impurities on the Forest Dislocation Density

11.2.1. Influence of the Divalent Impurities on the Forest Dislocation Density in Different Plastic Deformation Regions of Stress versus Strain Curve

Three-stage strain hardening has been clearly established as the characteristic behavior of single crystals of rock salt structure, deforming predominantly by single glide [1]. It is also observed for KCl (see figure 10-4 (c) in sec. 10.4 of chap. 10 or ref. [2, 3]) and KCl: Mg^{2+}, Ca^{2+}, Sr^{2+} or Ba^{2+} single crystals. The three-stage of KCl:Ba^{2+} is illustrated in figure 11-2, which is generally denoted by stage I, II, and III. The relation between temperature and $\Delta(\Delta\tau'/\Delta\ln\dot{\varepsilon})/\Delta\varepsilon$ in stage I and stage II of stress versus strain curve was investigated for the five kinds of specimens. The results are shown in figure 11-3 for KCl:Mg^{2+} and KCl, figure 11-4 for KCl:Ca^{2+} and KCl, figures 11-5 (a) and (b) for KCl:Sr^{2+} and KCl, and figure 11-6 for KCl:Ba^{2+} and KCl respectively. The full lines in these figures are to guide the reader's eye. Similar phenomena as described previously for KCl doped with monovalent impurities (see figure 10-6 in sec. 10.4 of chap. 10) are also observed for KCl doped with divalent additions. They are the following two phenomena. First, $\Delta(\Delta\tau'/\Delta\ln\dot{\varepsilon})/\Delta\varepsilon$ in stage II is larger than that in stage I at a given temperature for all of the specimens. This suggests that the forest dislocation density for the specimens slightly increases with the shear strain in stage I and then rapidly increases in stage II. Secondly, the $\Delta(\Delta\tau'/\Delta\ln\dot{\varepsilon})/\Delta\varepsilon$ in stage I and stage II increases with decreasing temperature for KCl and KCl doped with Ca^{2+}, Sr^{2+} or Ba^{2+}. From the second phenomenon, the forest dislocation density at a given shear strain in stage I and stage II may be considered to increase with decreasing temperature when the specimens are deformed in compression test. The dislocation density on the surface of Ge single crystals also increased with decreasing temperature in the same deformation region of stage I and stage II during the tensile test [4]. In figures 11-3 to 11-6, we can additionally observe that the value of $\Delta(\Delta\tau'/\Delta\ln\dot{\varepsilon})/\Delta\varepsilon$ for KCl doped with the divalent impurities is markedly larger as compared with that for KCl in stage I and stage II. As the divalent impurities in KCl increase $\Delta(\Delta\tau'/\Delta\ln\dot{\varepsilon})/\Delta\varepsilon$, it may be deduced that the forest dislocation density increases due to the divalent impurities. Dislocations multiply through the process of double cross-slip in LiF [5]. It may be predicted that the addition of divalent impurities influences the ability of cross slip or the frequency of cross slips. If it happens in KCl doped with the divalent impurities, the ability of dislocation multiplication will be changed by the impurities. When a dislocation in ionic crystal moves under an applied stress, the screw segments can cross slip, especially if they become held up by an obstacle [6]. The segments that then lie in the cross-slip plane have edge character and cannot slip conservatively in the direction of the Burgers vector. Hence these jogs on screw dislocations, which are produced by the cross slips, act as pinning points tending to restrain the forward motion of the screw segments [6]. The pinning points may act as dislocation sources. Consequently, the forest dislocation density would depend on the divalent additions. Etch pit technique revealed that there are screw dislocations on all possible active slip planes in a crystal of rock salt structure [7]. The ability of cross slip is generally governed also by the Peierls stress in the cross-slip plane. This, however, is not taken into account in this study, because the mobile dislocations in KCl are impeded by it below several tens K [8, 9].

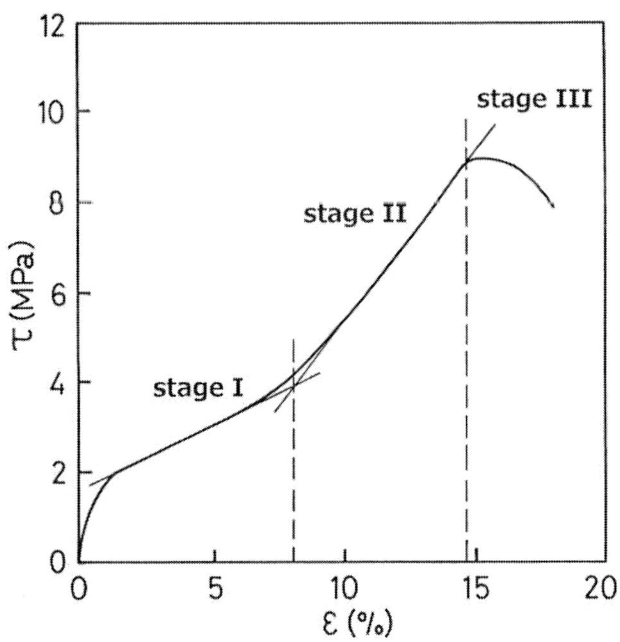

Figure 11-2. Stress versus strain curve for $KCl:Ba^{2+}$ (0.050 mol.%) at 145 K.

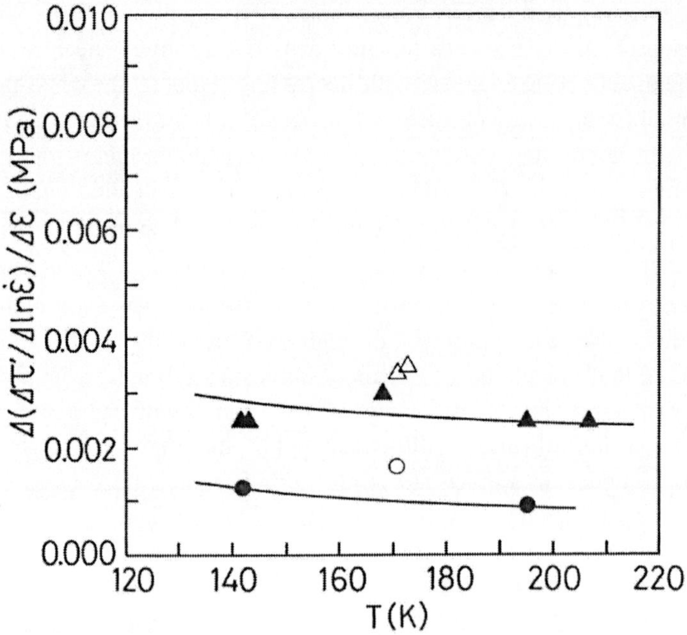

Figure 11-3. Dependence of $\Delta(\Delta\tau'/\Delta\ln\dot{\varepsilon})/\Delta\varepsilon$ on the temperature for KCl:Mg^{2+} (0.035 mol.%) and KCl in the different plastic deformation regions: (○) KCl:Mg^{2+} and (●) KCl in stage I; (Δ) KCl:Mg^{2+} and (▲) KCl in stage II.

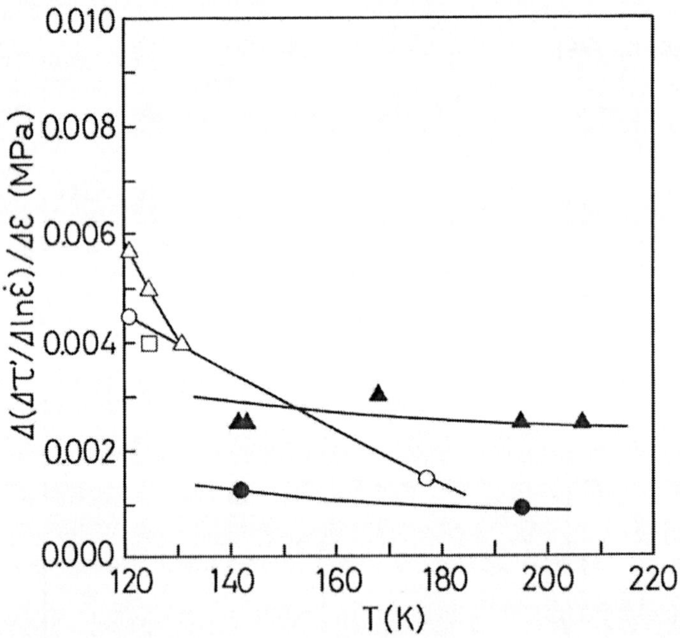

Figure 11-4. Dependence of $\Delta(\Delta\tau'/\Delta\ln\dot{\varepsilon})/\Delta\varepsilon$ on the temperature for KCl:Ca^{2+} and KCl in the different plastic deformation regions: (○) KCl:Ca^{2+} (0.035 mol.%), (□) KCl:Ca^{2+} (0.065 mol.%) and (●) KCl in stage I; (Δ) KCl:Ca^{2+} (0.035 mol.%) and (▲) KCl in stage II.

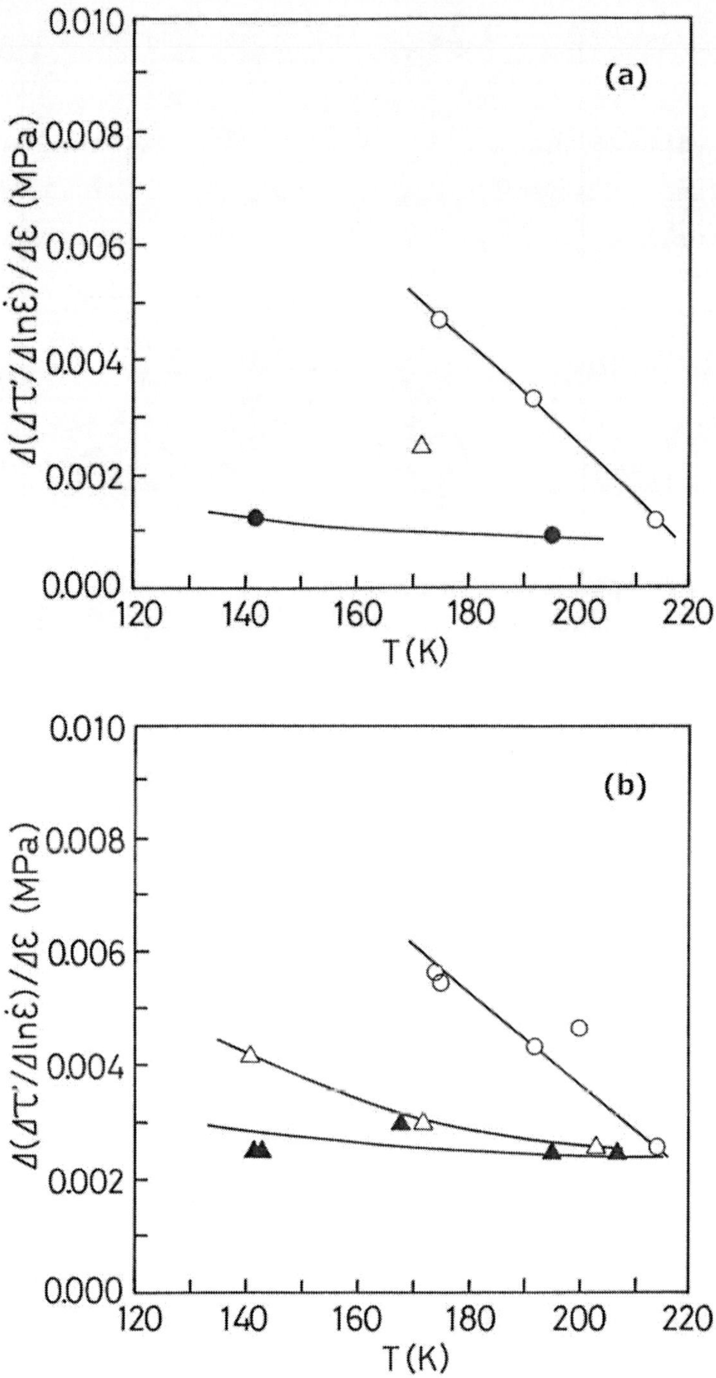

Figure 11-5. Dependence of $\Delta(\Delta\tau'/\Delta\ln\dot{\varepsilon})/\Delta\varepsilon$ on the temperature for $KCl:Sr^{2+}$ and KCl in the different plastic deformation regions: (a) (Δ) $KCl:Sr^{2+}$ (0.035 mol.%), (○) $KCl:Sr^{2+}$ (0.050 mol.%) and (●) KCl in stage I; (b) (Δ) $KCl:Sr^{2+}$ (0.035 mol.%), (○) $KCl:Sr^{2+}$ (0.050 mol.%) and (▲) KCl in stage II.

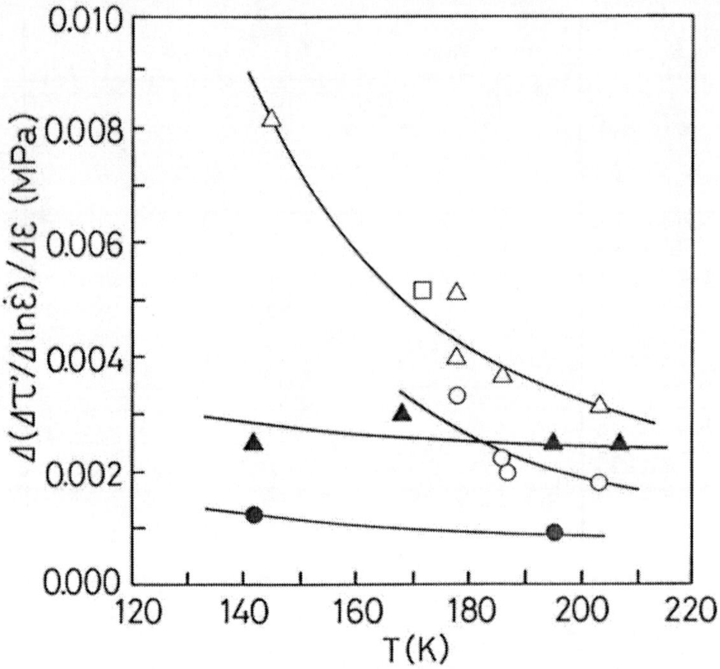

Figure 11-6. Dependence of $\Delta(\Delta\tau'/\Delta\ln\dot{\varepsilon})/\Delta\varepsilon$ on the temperature for KCl:Ba^{2+} (0.050 mol.%) and (●) KCl in stage I; (△) KCl:Ba^{2+} (0.050 mol.%), (□) KCl:Ba^{2+} (0.065 mol.%) and (▲) KCl in stage II.

The screw segments can cross slip with the aid of thermal fluctuations, where the activation volume is approximately 10 to 10^2 b^3 [10], so that the frequency of cross slips seems to be higher at high temperature. This may lead to multiply the forest dislocation density, i.e. $\Delta(\Delta\tau'/\Delta\ln\dot{\varepsilon})/\Delta\varepsilon$. But the results shown in figures 11-3 to 11-6 are contrary to the above expectation. It may be considered that the screw segments can more easily move forward by breaking away from the impurities on the slip plane, where the activation volume seems to be on the order of 10^2 b^3 [11], rather than the cross slips with increasing temperature.

11.2.2. Influence of the Concentration of Divalent Impurities on the Forest Dislocation Density

In figures 11-5 (a) and (b) for KCl:Sr^{2+} and figure 11-6 for KCl:Ba^{2+}, $\Delta(\Delta\tau'/\Delta\ln\dot{\varepsilon})/\Delta\varepsilon$ at a given temperature increases with the concentration of divalent impurities. That is, the forest dislocation density becomes high with the concentration of divalent impurities in the compression test. Similar result has been found for NaCl:Ca^{2+} single crystals at the beginning of deformation (ε=1 %, T=222.5 K) [12]. As can be seen from figures 11-5 (a) and (b), this is clear in the two stages for KCl:Sr^{2+} and further becomes more clearly at low temperature. The mobility of jogs on the screw dislocations, where the activation volume for non-conservative motion of jogs is approximately 10^2 to 10^4 b^3 [10], would become low with the addition of divalent impurities. This may give rise to multiply jogs on the screw dislocations with increasing the impurity concentration. Therefore, the dislocation density seems to be higher with the concentration in accordance with the dislocation behavior mentioned in the preceding section 11.2.1. Unfortunately, $\Delta(\Delta\tau'/\Delta\ln\dot{\varepsilon})/\Delta\varepsilon$ in stage I for KCl:Ba^{2+} (0.065 mol.% in the melt) could not be derived on the basis of strain-rate sensitivity versus stress decrement curves. In figure 11-4, no difference between $\Delta(\Delta\tau'/\Delta\ln\dot{\varepsilon})/\Delta\varepsilon$ for KCl:Ca^{2+} (0.035 mol.% in the melt) and for KCl:Ca^{2+} (0.065 mol.% in the melt) can be observed in stage I at 125 K. This is because the concentration of divalent impurities (*c*) in KCl:Ca^{2+} (0.035 mol.% in the melt), of which *c* is determined to be 43.1 ppm, is nearly equal to that for KCl:Ca^{2+} (0.065 mol.% in the melt), of which *c* is 43.5 ppm.

11.2.3. Influence of the Size of Divalent Impurity on the Forest Dislocation Density

It was investigated whether the forest dislocation density is influenced by a change in the size of divalent impurity in the specimens. Figures 11-7 (a) to (c) show the $\Delta(\Delta\tau'/\Delta\ln\dot{\varepsilon})/\Delta\varepsilon$ at almost the same concentration of divalent impurities around 171 K. The $\Delta(\Delta\tau'/\Delta\ln\dot{\varepsilon})/\Delta\varepsilon$ in figures 11-7 (a) and (b) is in stage I and that in figure 11-7 (c) is in stage II. $\Delta(\Delta\tau'/\Delta\ln\dot{\varepsilon})/\Delta\varepsilon$, i.e. the forest dislocation density, in the two stages increases with the size of divalent addition. Therefore, it is considered that the frequency of cross slips grow higher when the divalent cationic size approaches to the K^+ ionic size at the temperature.

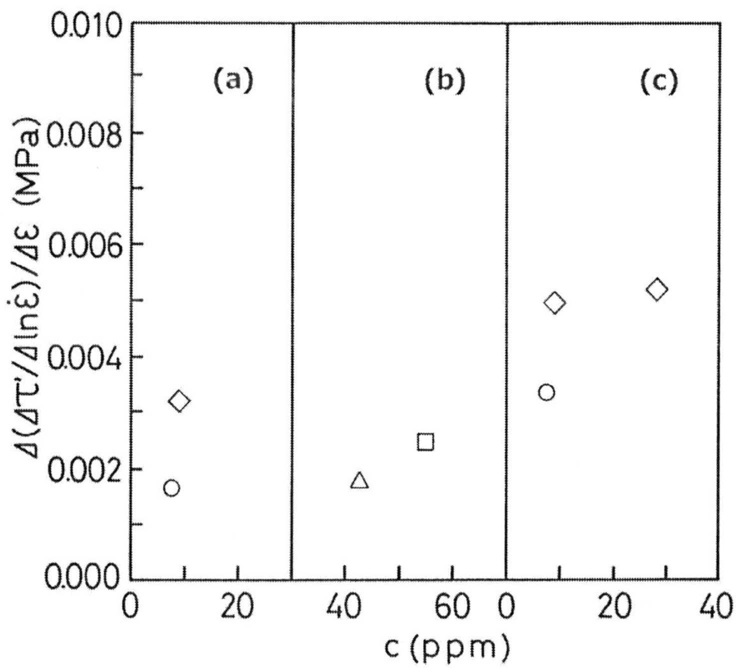

Figure 11-7. $\Delta(\Delta\tau'/\Delta\ln\dot{\varepsilon})/\Delta\varepsilon$ for (○) KCl:Mg^{2+}, (Δ) KCl:Ca^{2+}, (□) KCl:Sr^{2+} and (◇) KCl:Ba^{2+} at various conditions: (a) stage I and 171 K, (b) stage I and 172 K and (c) stage II and 171 K. The concentration of divalent impurities, *c*, is obtained by dielectric loss measurement.

11.3. Influence of the Divalent Impurities on Mobile Dislocation Density

In this section, it is described whether the difference in size of divalent impurity influences mobile dislocation density for KCl doped with Ca^{2+}, Sr^{2+} or Ba^{2+}. As the mobile dislocation density for bulk cannot be determined directly, the density was investigated from pre-exponential factor ($\dot{\varepsilon}_0$) by reason of the description [12]: the most probable variable in $\dot{\varepsilon}_0$ is due to the density of mobile dislocations.

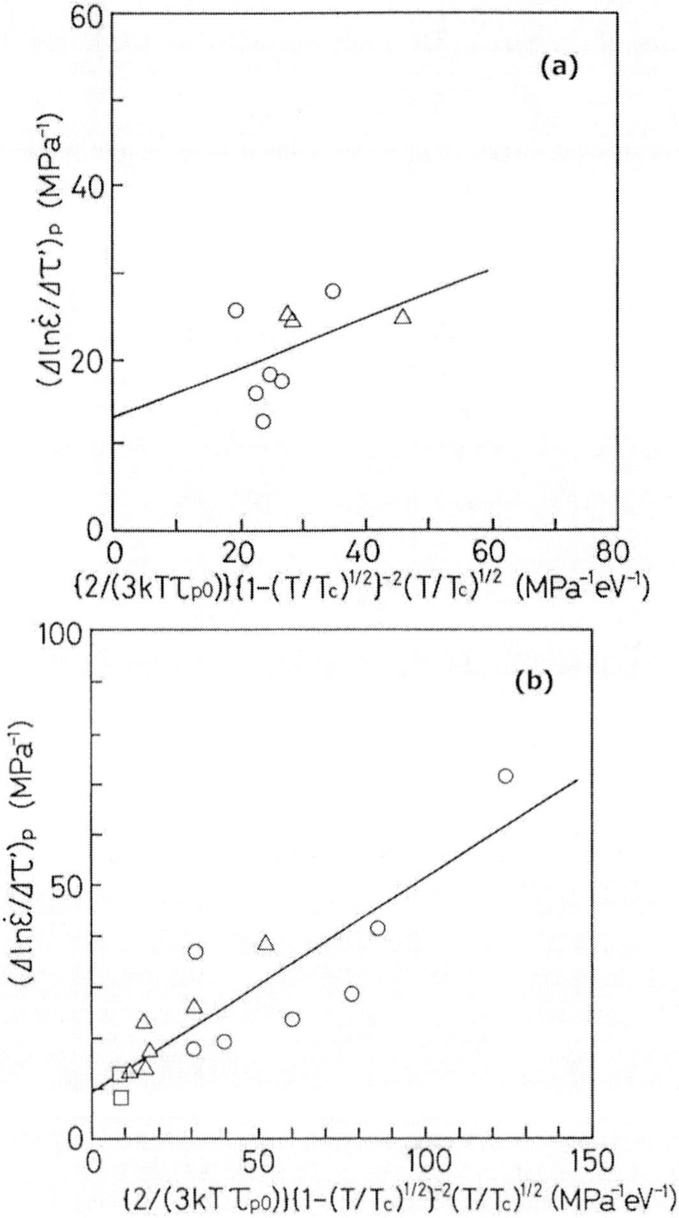

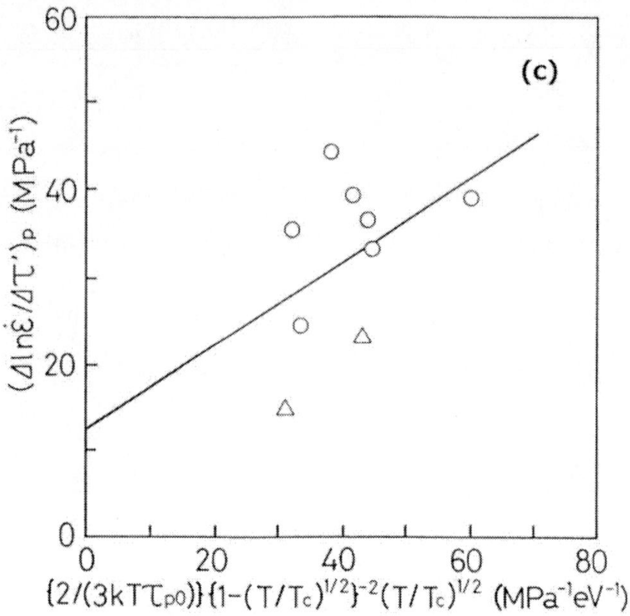

Figure 11-8. Linear plots of equation (11-1) for (a) KCl:Ca^{2+} (○) 0.035 mol.% and (Δ) 0.065 mol.%, for (b) KCl:Sr^{2+} (○) 0.035 mol.%, (Δ) 0.050 mol.% and (□) 0.065 mol.% and for (c) KCl:Ba^{2+} (○) 0.050 mol.% and (Δ) 0.065 mol.%.

If the interaction between a dislocation and an impurity can be approximated to the Fleischer's model [13] taking account of the Friedel relation [14], the Gibbs free energy, ΔG_0, for the breakaway of the dislocation from the impurity is calculated from the following equation (see sec. 5.6 of chap. 5):

$$\frac{\partial \ln \dot{\varepsilon}}{\partial \tau^*} = \left(\frac{2\Delta G_0}{3kT\tau_0^*}\right)\left\{1-\left(\frac{T}{T_c}\right)^{1/2}\right\}^{-2}\left(\frac{T}{T_c}\right)^{1/2} + \frac{\partial \ln \dot{\varepsilon}_0}{\partial \tau^*}, \tag{11-1}$$

where kT has the usual meaning, τ^*_0 is the effective shear stress due to the impurities at 0 K, T_c is the critical temperature T, at which the effective shear stress is zero. The model is hereafter termed the F-F. The results of equation (11-1) are shown in figure 11-8 (a) for KCl:Ca^{2+}, (b) for KCl:Sr^{2+}, and (c) for KCl:Ba^{2+} respectively. In figures 11-8 (a) to (c), τ^*_0 is replaced by τ_{p0}. τ_{p0} is the effective shear stress due to impurities without thermal activation (see sec. 4.2 and 4.3 of chap. 4) and is re-tabulated in table 11-1. In addition, the $\partial \ln \dot{\varepsilon}/\partial \tau^*$ in equation (11-1) is represented by $(\Delta \ln \dot{\varepsilon}/\Delta \tau')_p$. The $(\Delta \tau'/\Delta \ln \dot{\varepsilon})_p$, which is given by the difference between strain-rate sensitivity at first plateau place and at second one on the strain-rate sensitivity versus stress decrement curve as presented in figure 11-1, is assumed to be the strain-rate sensitivity due to impurities (see sec. 4.4 of chap. 4). Accordingly, the $\partial \ln \dot{\varepsilon}_0/\partial \tau^*$ in equation (11-1) can be represented by $\Delta \ln \dot{\varepsilon}_0/\Delta \tau'$. Then, the values of $\Delta \ln \dot{\varepsilon}_0/\Delta \tau'$, at which the solid lines intersect the ordinate in figures 11-8 (a) to (c), are given in table 11-2. $\dot{\varepsilon}_0$ is expressed by [15, 16]

$$\dot{\varepsilon}_0 = \rho b^2 \nu_D (L_0/L)^2, \tag{11-2}$$

where is the density of mobile dislocations, *b* is the magnitude of the Burgers vector, is the Debye frequency, is the average spacing of impurities on the slip plane and is the average length of dislocation segments at 0 K. The may be given by the Friedel relation [14]:

$$L = \left(\frac{2L_0^2 E}{\tau_0^* b} \right)^{1/3}, \tag{11-3}$$

where *E* is the line tension of the dislocations and is calculated by μb^2. is the shear modulus at 0 K. Substituting equation (11-3) in equation (11-2), we find

$$\dot{\varepsilon}_0 = \rho \nu_D \left\{ \frac{b^8 L_0^{\ 2} \tau_0^{*2}}{(2E)^2} \right\}^{1/3}. \tag{11-4}$$

Differentiating the natural logarithmic equation of equation (11-4) with respect to the effective shear stress gives

$$\frac{\partial \ln \dot{\varepsilon}_0}{\partial \tau^*} = \frac{\partial \ln \rho}{\partial \tau^*} + \frac{2}{3\tau_{p0}}, \tag{11-5}$$

where τ^*_0 is replaced by τ_{p0}. Substituting the values of $\Delta \ln \dot{\varepsilon}_0 / \Delta \tau'$ and τ_{p0} into equation (11-5), $(\Delta \ln \rho / \Delta \tau')$ can be obtained for the three kinds of specimens. The results are listed in table 11-3. The values of $(\Delta \rho / \Delta \tau')$ are within stage II of stress versus strain curve, since almost all of the $(\Delta \ln \dot{\varepsilon} / \Delta \tau')_p$ in figures 11-8 (a) to (c) are obtained on the basis of the strain-rate sensitivity versus stress decrement curve at the shear strain in stage II. However, $(\Delta \ln \rho / \Delta \tau')$ for $KCl{:}Mg^{2+}$ could not be obtained according to the above-mentioned process, because the F-F seemed not to be appropriate for the interaction between a dislocation and the impurity in $KCl{:}Mg^{2+}$ (see sec. 5.4 of chap. 5). As given in table 11-3, no great difference of $(\Delta \rho / \Delta \tau')$ between the three kinds of specimens can be unfortunately discriminated on account of scattered the experimental data in figures 11-8 (a) to (c). The values of $(\Delta \rho / \Delta \tau')$ for the specimens, however, seem to be less than about 50 cm^{-2} MPa^{-1}.

Table 11-1. Values of τ_{p0} for the F-F

Single crystal	τ_{p0} (MPa)
$KCl:Ca^{2+}$ (0.035 mol.%)	15.34[a]
(0.065 mol.%)	14.11[a]
$KCl:Sr^{2+}$ (0.035 mol.%)	11.44
(0.050 mol.%)	25.47
(0.065 mol.%)	36.31
$KCl:Ba^{2+}$ (0.050 mol.%)	5.34
(0.065 mol.%)	7.27

[a] No difference between τ_{p0} can be observed, because the concentration of divalent cations in $KCl:Ca^{2+}$ (0.035 mol.%) is nearly equal to that for $KCl:Ca^{2+}$ (0.065 mol.%) (see table 5-3 on page 77).

Table 11-2. Values of $\Delta\ln\dot{\varepsilon}_0/\Delta\tau'$

Single crystal	$\Delta\ln\dot{\varepsilon}_0/\Delta\tau'$ (MPa^{-1})
$KCl:Ca^{2+}$	13.1
$KCl:Sr^{2+}$	9.0
$KCl:Ba^{2+}$	12.5

Table 11-3. Values of $(\Delta\rho/\Delta\tau')$ at 0 K

Single crystal	$(\Delta\rho/\Delta\tau')$ (cm^{-2} MPa^{-1})
$KCl:Ca^{2+}$ (0.035 mol.%)	46.84
(0.065 mol.%)	46.65
$KCl:Sr^{2+}$ (0.035 mol.%)	0.76
(0.050 mol.%)	0.79
(0.065 mol.%)	0.80
$KCl:Ba^{2+}$ (0.050 mol.%)	23.68
(0.065 mol.%)	24.48

CONCLUSION

When the specimens are deformed in compression tests, the following five phenomena are found on the basis of variation of strain-rate sensitivity at the second plateau place on the curve with shear strain (i.e. $\Delta(\Delta\tau'/\Delta\ln\dot{\varepsilon})/\Delta\varepsilon$), which is considered to represent a change in forest dislocation density with shear strain.

1. The forest dislocation density for the specimens slightly increases with the shear strain in stage I and then rapidly increases in stage II.
2. The forest dislocation density increases with decreasing temperature at a given shear strain in stage I and stage II.
3. In the two stages, the forest dislocation density increases due to the divalent impurities in the specimens. It may be predicted that the addition of divalent

impurities influences the ability of cross slip or the frequency of cross slips. We can image that the screw segments can cross slip under an applied stress, especially if they become held up by the impurity. These jogs on screw dislocations, which are produced by the cross slips, seem to act as pinning points tending to restrain the forward motion of the screw segments and further as dislocation sources. Consequently, the forest dislocation density would depend on the divalent additions.

4. The forest dislocation density becomes high with the concentration of the divalent impurities which may result in a large number of jogs on the screw dislocations. This becomes more clearly at low temperature.
5. When the divalent cationic size approaches to the K^+ ionic size (see figure 4-11 in sec. 4.3 of chap. 4), the forest dislocation density increases around 171 K. This suggests that the frequency of cross slips grow higher with larger size of divalent impurity.

Unfortunately, it was not possible to determine whether the change in the size of divalent impurity influences mobile dislocation density from the values of $(\Delta\rho/\Delta\tau')$ for KCl doped with Ca^{2+}, Sr^{2+} or Ba^{2+} by reason of scattered the experimental data in figures 11-8 (a) to (c). The values of $(\Delta\rho/\Delta\tau')$ for the specimens in stage II, however, seems to be less than about 50 cm^{-2} MPa^{-1} in the compression tests.

REFERENCES

[1] Sprackling, M. T. *The Plastic Deformation of Simple Ionic Crystals*, (edited by Alper, A. M.; Margrave, J. L.; Nowick, A. S.) Academic Press; London, 1976; p 203.
[2] Alden, T. H. *Trans. Met. Soc. AIME* 1964, *230*, 649.
[3] Davis, L. A.; Gordon, R. B. *J. Appl. Phys.* 1969, *40*, 4507.
[4] Kojima, K.; Sumino, K. *Cryst. Latt. Defects* 1971, *2*, 147.
[5] Gilman, J. J.; Johnston, W. G. *Solid State Physics* 1962, *13*, 147.
[6] Sprackling, M. T. *The Plastic Deformation of Simple Ionic Crystals*, (edited by Alper, A. M.; Margrave, J. L.; Nowick, A. S.) Academic Press; London, 1976; p 109.
[7] Sprackling, M. T. *The Plastic Deformation of Simple Ionic Crystals*, (edited by Alper, A. M.; Margrave, J. L.; Nowick, A. S.) Academic Press; London, 1976; p 195.
[8] Kataoka, T.; Uematsu, T.; Yamada, T. *Jpn. J. Appl. Phys.* 1978, *17*, 271.
[9] Suzuki, T. *Oyo Buturi* 1976, *45*, 447 (in Japanese).
[10] Conrad, H. *J. Metals* 1964, *16*, 582.
[11] Srinivasan, M.; Stoebe, T. G. *J. Mater. Sci.* 1973, *8*, 1567.
[12] Appel, F. *Phys. Status Solidi* (a) 1980, *61*, 477.
[13] Fleischer, R. L. *J. Appl. Phys.* 1962, *33*, 3504.
[14] Friedel, J. *Dislocations;* Pergamon Press; Oxford, 1964; p 224.
[15] Sumino, K. *Plasticity of Crystals,* Maruzen; Tokyo, 1977; p 356 (in Japanese).
[16] Sumino, K. *Jpn. Inst. Met.* 1972, *11*, 31 (in Japanese).

Chapter 12

INFLUENCE OF THE HEAT TREATMENT ON DEFORMATION CHARACTERISTICS

12.1. INTRODUCTION

When alkali halide crystals are doped with divalent ions, the ions are expected to be paired with positive ion vacancies. The pairs are termed I-V dipoles. If the I-V dipoles in KCl:Sr^{2+} single crystals aggregate by heat treatment, the change in the state of a small amount of impurities (I-V dipoles) strongly influences the resistance to movement of a dislocation [1]. Therefore, various deformation characteristics will be changed. For example, the Fleischer's model [2] taking account of the Friedel relation [3] is suitable for the interaction between a dislocation and the impurity in KCl:Sr^{2+} crystals but not for the annealed ones (see sec. 5.5 of chap. 5 and sec. 7.2.2 of chap. 7). This was examined mainly on the basis of the dependence of strain-rate sensitivity due to the impurities on temperature. In chapter 7, the influence of the heat treatment on the various characteristics (the force acted on the dislocation, the bending angle of dislocation by the impurity, dislocation velocity-effective stress exponent, etc.) is described from the relative curve of strain-rate sensitivity and stress decrement due to ultrasonic oscillation. The curve is considered to reflect the influence of ultrasonic oscillation on the dislocation motion on the slip plane containing many impurities and a few forest dislocations (see sec. 2.7 of chap. 2). In this chapter 12, it is mainly investigated whether the density of the forest dislocation is changed by the heat treatment under the compression test. With respect to the two kinds of KCl:Sr^{2+} (0.05 mol.% in the melt) single crystals (i.e., the quenched specimen and the annealed one) described in the section 3.3 of chapter 3, this is examined from the variation of strain-rate sensitivity at the second plateau place on the curve with shear strain carried out in the preceding chapters: sec. 10.4 of chap. 10 and sec. 11.2 of chap. 11.

12.2. INFLUENCE OF THE HEAT TREATMENT ON MOVING DISLOCATION DENSITY

When the thermally activated overcoming of the aggregates controls the dislocation velocity, the strain rate ($\dot{\varepsilon}$) is expressed by an Arrhenius-type equation [4]

$$\dot{\varepsilon} = \dot{\varepsilon}_0 \exp\left(\frac{-\Delta G}{kT}\right), \tag{12-1}$$

where $\dot{\varepsilon}_0$ is a frequency factor, is the change in Gibbs free energy of activation for the dislocation motion and kT has the usual meaning. The force (F) acting on the dislocation as a function of the distance (x) until the aggregate on the slip plane is given by

$$F(x) = F_0(1 - |x|^n / a^n),\ |x| < a \tag{12-2}$$

$$(F(\pm a) = 0,\ F(0) = F_0),$$

where square is for the case of $n = \infty$, triangular for $n = 1$ and parabolic for $n = 2$ [5]. A square force-distance relation, a parabolic one and a triangular one are termed the SQ, the PA and the TR, respectively, in this chapter. Integrating equation (12-2) with respect to x, the Gibbs free energy is given by [6]

$$\Delta G = 2F_0 a\left(\frac{n}{n+1}\right)\left(1 - \frac{F}{F_0}\right)^{(n+1)/n}. \tag{12-3}$$

The Gibbs free energy for the SQ is obtained from the Friedel relation [3]:

$$L = \left(\frac{2L_0^{\,2}E}{\tau^* b}\right)^{1/3} \tag{12-4}$$

namely,

$$\Delta G = 2F_0 a\left(1 - \frac{F}{F_0}\right), \tag{12-5}$$

where is the average spacing of impurities on the slip plane, E is the line tension of the dislocations, τ^* is the effective stress and b is the magnitude of the Burgers vector. F and F_0 in equation (12-5) are expressed by using equation (12-4) as follows

$$F = \left(2L_0^{\,2}E\right)^{1/3} b^{2/3} \tau^{*2/3} \tag{12-6}$$

and

$$F_0 = \left(2L_0^{\,2}E\right)^{1/3} b^{2/3} \tau_0^{*2/3}, \tag{12-7}$$

where τ^*_0 is the effective stress at 0 K. When the F is zero, equation (12-3) can be transformed as the following:

$$\Delta G_0 = 2F_0 a\left(\frac{n}{n+1}\right). \qquad (12\text{-}8)$$

Substituting of equations (12-6) to (12-8) in equation (12-5), the Gibbs free energy for the SQ yields

$$\Delta G = \Delta G_0\left\{1-\left(\frac{\tau^*}{\tau^*_0}\right)^{2/3}\right\}. \qquad (12\text{-}9)$$

Differentiating the substitutional equation of equation (12-9) in equation (12-1) with respect to the shear stress, we find

$$\frac{\partial \ln \dot{\varepsilon}}{\partial \tau^*} = \left(\frac{2\Delta G_0}{3k}\right)\left(\tau_{p0}{}^2\tau_{p1}\right)^{-1/3} \Big/ T + \frac{\partial \ln \dot{\varepsilon}_0}{\partial \tau^*}, \qquad (12\text{-}10)$$

where τ^*_0 and τ^* are replaced by τ_{p0} and τ_{p1}. τ_{p1} is considered to be the effective shear stress due to only one type of the impurities (see sec. 2.7 of chap. 2) and τ_{p0} that due to the impurities without thermal activation (see sec. 4.3 of chap. 4) when the dislocation moves forward with the help of oscillation. $\partial \ln \dot{\varepsilon}/\partial \tau^*$ for the PA is expressed by (see sec. 7.3 of chap. 7)

$$\frac{\partial \ln \dot{\varepsilon}}{\partial \tau^*} = \frac{\Delta G_0}{kT}\left\{\left(\tau_{p0}^2\tau_{p1}\right)^{-2/3} - \tau_{p0}^{-2}\right\}^{1/2} + \frac{\partial \ln \dot{\varepsilon}_0}{\partial \tau^*} \qquad (12\text{-}11)$$

and that for the TR yields

$$\frac{\partial \ln \dot{\varepsilon}}{\partial \tau^*} = \frac{4\Delta G_0}{3kT}\left\{\left(\tau_{p0}^2\tau_{p1}\right)^{-1/3} - \left(\tau_{p0}^{-4}\tau_{p1}\right)^{1/3}\right\} + \frac{\partial \ln \dot{\varepsilon}_0}{\partial \tau^*}, \qquad (12\text{-}12)$$

where τ_{p0} values for the three kinds of force-distance relations are re-tabulated in table 12-1 (see sec. 7.2.1 of chap. 7). The results of calculations for equations (12-10) to (12-12) are shown in figures 12-1 (a) to (c) in which the $\partial \ln \dot{\varepsilon}/\partial \tau^*$ is represented by $(\Delta \ln \dot{\varepsilon} / \Delta \tau')_p$.

The $(\Delta \tau' / \Delta \ln \dot{\varepsilon})_p$, which is given by the difference between strain-rate sensitivity at first plateau place and at second one on the relative curve of strain-rate sensitivity and stress decrement, is assumed to be the strain-rate sensitivity due to the impurities (see sec. 4.4 of chap. 4).

Accordingly, $\partial \ln \dot{\varepsilon}_0 / \partial \tau^*$ in equations (12-10) to (12-12) can be represented by $\Delta \ln \dot{\varepsilon}_0 / \Delta \tau'$. The $\Delta \ln \dot{\varepsilon}_0 / \Delta \tau'$ corresponds to the value at which the solid lines intersect the ordinate in these figures.

The results are given in table 12-1. The $\Delta \ln \dot{\varepsilon}_0 / \Delta \tau'$ is expressed by (see sec. 11.3 of chap. 11)

$$\frac{\Delta \ln \dot{\varepsilon}_0}{\Delta \tau'} = \frac{\Delta \ln \rho}{\Delta \tau'} + \frac{2}{3\tau_{p0}}, \tag{12-13}$$

where is the density of mobile dislocation. Substituting the values of $\Delta \ln \dot{\varepsilon}_0 / \Delta \tau'$ and τ_{p0} into equation (12-13), $(\Delta \rho / \Delta \tau')$ can be obtained.

These values for the three force-distance relations are given in table 12-2. The values of $(\Delta \rho / \Delta \tau')$ are within stage II of stress-strain curve, because the $(\Delta \tau' / \Delta \ln \dot{\varepsilon})_p$ values in figures 12-1 (a) to (c) are obtained on the basis of the relative curve of strain-rate sensitivity and stress decrement in stage II. Although the values of $(\Delta \ln \dot{\varepsilon} / \Delta \tau')_p$ are slightly scattering as shown in figures 12-1 (a) to (c), $(\Delta \rho / \Delta \tau')$ for the annealed specimen seems to be less than 3.38 cm^{-2} MPa^{-1}.

From table 12-2, it is difficult to discriminate the difference in $(\Delta \rho / \Delta \tau')$ among the three. On the other hand, that for the quenched specimen is 0.79 cm^{-2} MPa^{-1} (see sec. 11.3 of chap. 11). Therefore, it is considered that the density of moving dislocation is not influenced by the heat treatment in the strain-rate cycling tests under superposition of ultrasonic oscillatory shear stress.

Table 12-1. Values of τ_{p0} and $\Delta \ln \dot{\varepsilon}_0 / \Delta \tau'$ for the three force-distance relations between a dislocation and the aggregate in the annealed specimen

Force-distance relation	τ_{p0} (MPa)	$\Delta \ln \dot{\varepsilon}_0 / \Delta \tau'$ (MPa^{-1})
SQ	2.17	6.80
PA	3.28	10.46
TR	4.53	10.57

Table 12-2. Values of $\Delta \rho / \Delta \tau'$ for the three force-distance relations between a dislocation and the aggregate in the annealed specimen

Force-distance relation	$\Delta \rho / \Delta \tau'$ (cm^{-2} MPa^{-1})
SQ	0.07
PA	2.85
TR	3.38

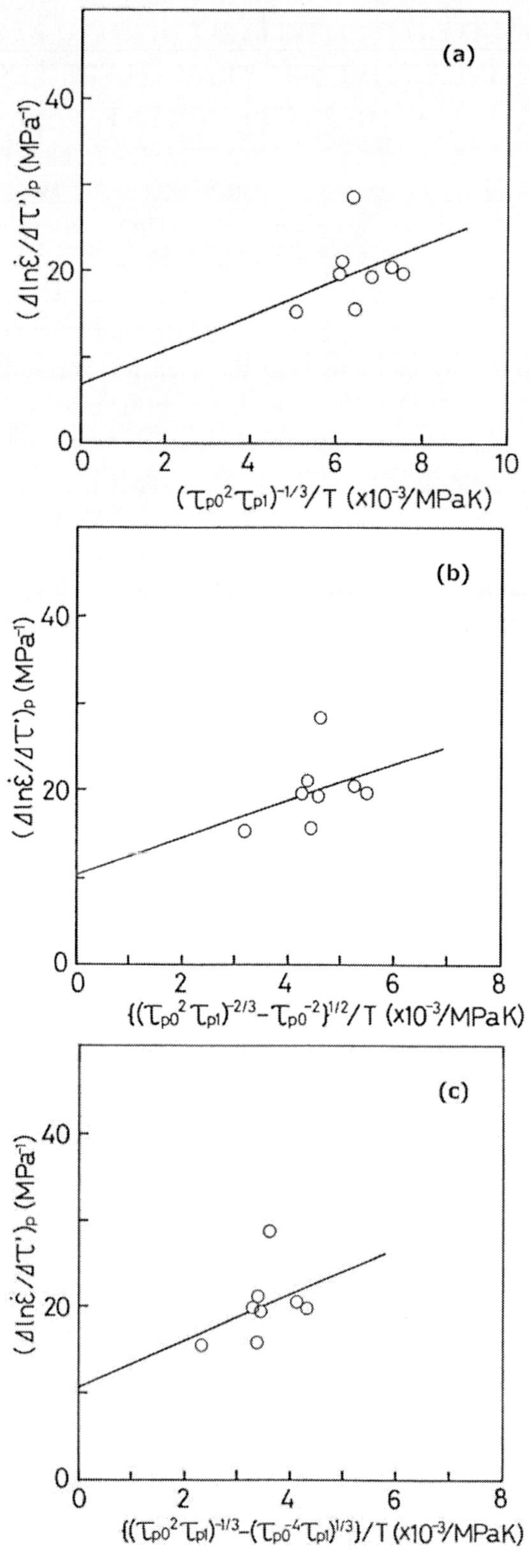

Figure 12-1. Linear plots of (a) equation (12-10) for the SQ, (b) equation (12-11) for the PA and (c) equation (12-12) for the TR. (○) : $(\Delta \ln \dot{\varepsilon} / \Delta \tau')_p$ for the annealed specimen.

12.3. STRAIN-RATE SENSITIVITY AT SECOND PLATEAU PLACE ON THE RELATIVE CURVE OF STRAIN-RATE SENSITIVITY AND STRESS DECREMENT

The strain-rate sensitivity at second plateau place on the strain-rate sensitivity versus stress decrement curve is considered to be due to dislocation cuttings (see sec. 2.7 of chap. 2). We investigate the relation between temperature and the variation of strain-rate sensitivity at the second plateau place on the relative curve of strain-rate sensitivity and stress decrement with shear strain, $\Delta(\Delta\tau'/\Delta\ln\dot{\varepsilon})/\Delta\varepsilon$, in the different plastic deformation regions of stress-strain curve for the two kinds of specimens, namely, the quenched specimen and the annealed one in the section 3.3 of chapter 3. The $\Delta(\Delta\tau'/\Delta\ln\dot{\varepsilon})/\Delta\varepsilon$ seems to be the variation of the strain-rate sensitivity due to dislocation cuttings with shear strain (see sec. 10.4 of chap. 10). Three-stage strain hardening is obtained for KCl [7, 8] and is also observed for KCl:Sr^{2+} (0.05 mol.% in the melt). The size of specimens used at the following research in this section is within (5±1)×(5±1)×(15±1) mm^3. Because, it has been reported that the height-width of specimens determines the shape of the work-hardening curve in the compression at a constant strain rate and temperature for CaF_2 single crystals [9] and besides glide geometry is directly related to the shape and size of KCl crystals [10].

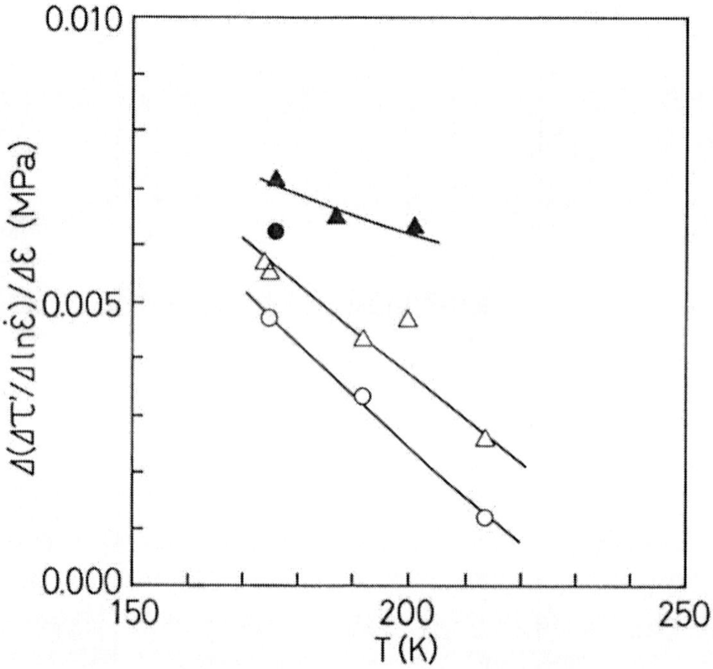

Figure 12-2. Dependence of $\Delta(\Delta\tau'/\Delta\ln\dot{\varepsilon})/\Delta\varepsilon$ on the temperature in the different plastic deformation regions: (○) for the quenched specimen and (●) for the annealed specimen in stage I; (Δ) for the quenched specimen and (▲) for the annealed specimen in stage II.

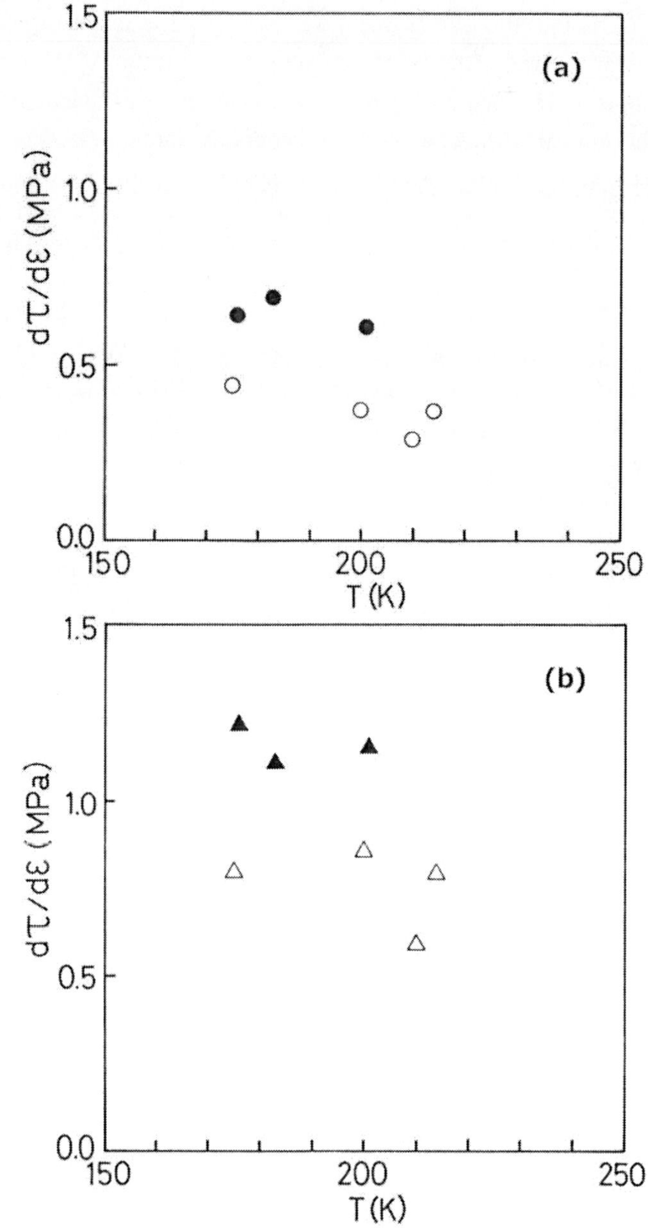

Figure 12-3. Dependence of the strain-hardening rate on the temperature in the different plastic deformation regions: (a) (○) for the quenched specimen and (●) for the annealed specimen in stage I; (b) (Δ) for the quenched specimen and (▲) for the annealed specimen in stage II.

In figure 12-2, the relations of temperature and $\Delta(\Delta\tau'/\Delta\ln\dot{\varepsilon})/\Delta\varepsilon$ in stage I and in stage II are represented by a circle and a triangle, respectively. The open symbols correspond to that for the quenched specimen and the solid ones that for the annealed specimen. The curves in figure 12-2 are to guide the reader's eye. Unfortunately, the $\Delta(\Delta\tau'/\Delta\ln\dot{\varepsilon})/\Delta\varepsilon$ could not be obtained at low temperature. The same two phenomena as described for KCl:Li^+ (0.50 mol.% in the melt) and KCl:Na^+ (0.50 mol.% in the melt) in the chapter 10 (see sec. 10.4) are also observed for both the quenched specimen and the annealed specimen in figure 12-2. That is, the first phenomenon is that the $\Delta(\Delta\tau'/\Delta\ln\dot{\varepsilon})/\Delta\varepsilon$ in stage II is obviously larger than that in stage I at a given temperature. The other phenomenon is that the $\Delta(\Delta\tau'/\Delta\ln\dot{\varepsilon})/\Delta\varepsilon$ in stage I and in stage II increases with decreasing temperature. Figure 12-2 also shows that the $\Delta(\Delta\tau'/\Delta\ln\dot{\varepsilon})/\Delta\varepsilon$ for the annealed specimens is considerably large in contrast to that for the quenched specimens in the two stages at the temperature. This may result from a rapid increase in forest dislocation density with shear strain in the annealed specimen. Accordingly, the increase in that in the annealed specimen seems to be remarkable in the two stages under the compression test, compared with that in the quenched specimen. According to Michalak [11] some dislocation cutting process becomes an important hardening mechanism and the strain-hardening increases. If the consideration about forest dislocation density for the two kinds of specimens is accurate, it is expected that a strain-hardening rate ($d\tau/d\varepsilon$) for the annealed specimen is larger than that for the quenched specimen at a given temperature and shear strain.

When the whole crystal is filled with primary dislocation dipoles and multipoles, stage I ends and the stage II, which is characterized by the onset of glide on oblique systems, follows. Afterwards, the stage III, which is probably associated with the cross glide of screw dislocation around obstacles, ensues in the process of plastic deformation [12]. Accordingly, it will be also expected that the extent of stage I and stage II for the annealed specimen is shorter than that for the quenched specimen at a given temperature.

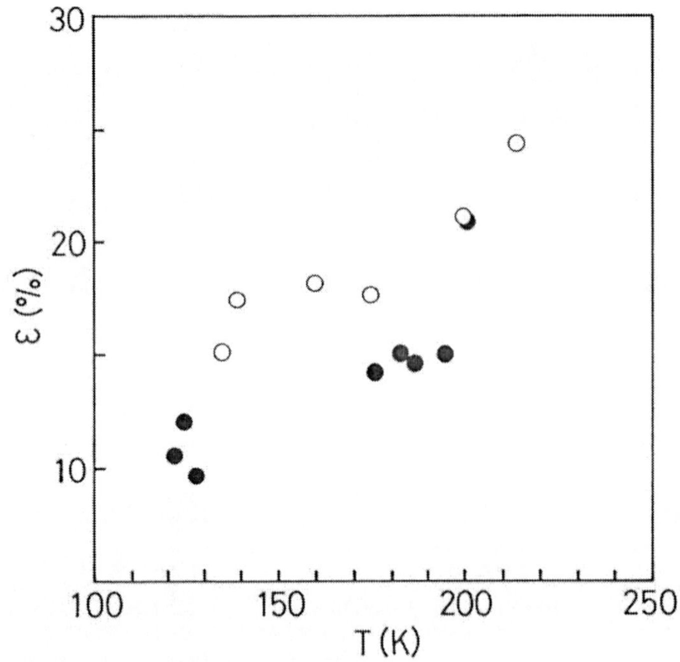

Figure 12-4. Dependence of the extent of plastic deformation region on the temperature for both the specimens: (○) for the quenched specimen and (●) for the annealed specimen.

The relation between strain-hardening rate and temperature for the two kinds of specimens is shown in figure 12-3 (a) for stage I and (b) for stage II. The value of strain-hardening rate in stage II is about two times larger than that in stage I at a given temperature for both the specimens. Furthermore, the strain-hardening rate for the annealed specimen is obviously large as against that for the quenched specimen in the two stages as can be seen from figures 12-3 (a) and (b). We then investigate the extent of stage I and stage II. However, the difference between the extents of stage I for the two kinds of specimens could not be found out within the temperature. The extent of stage II for the quenched specimen could not also be compared with that for the annealed specimen, because most of the specimens were fractured within stage II by the compression test. Then, we attempt to investigate the extent between shear strains of yield point and of fractured point, i.e. plastic deformation region, at various temperatures. The result of the extent of plastic deformation region is shown in figure 12-4. The open circles and the solid ones correspond to that for the quenched specimen and for the annealed specimen, respectively. The extent of it tends to be large with increasing temperature for both the specimens. Figure 12-4 also shows that the extent of that for the annealed specimen seems to be shorter than that for the quenched specimen at a given temperature. Consequently, it is clear that the increase in forest dislocation density in the annealed specimen is remarkable in contrast to that in the quenched specimen under the compression test. This may be attributed to the phenomenon that a small amount of the much larger aggregates than trimers (see figure 3-5 in sec. 3.3 of chap. 3), which are contained in the annealed specimen, contribute to the increase in forest dislocation density.

CONCLUSION

The following conclusions are derived from the data obtained on the basis of the relative curve between strain-rate sensitivity and stress decrement due to the oscillation.

1. It is difficult to discriminate the difference of $(\Delta\rho/\Delta\tau')$ among the three force-distance relations (the SQ, the PA, the TR) because of the scattering of $(\Delta\ln\dot{\varepsilon}/\Delta\tau')_p$. However, the increase in mobile dislocation density with the stress change, $(\Delta\rho/\Delta\tau')$, for the annealed specimen seems to be less than 3.38 cm^{-2} MPa^{-1}. This suggests that the density of moving dislocation is not influenced by the heat treatment in this chapter.
2. The variation of strain-rate sensitivity at the second plateau place on the curve with shear strain, $\Delta(\Delta\tau'/\Delta\ln\dot{\varepsilon})/\Delta\varepsilon$, for the annealed specimen is considerably large in contrast to that for the quenched specimen in both stage I and stage II of stress-strain curve at 170 to 220 K. Accordingly, the increase in forest dislocation density with shear strain for the annealed specimen seems to be remarkable in the two stages under the compression test, compared with that for the quenched specimen. As a proof of it, the following two experimental results are observed. First, the strain-hardening rate for the annealed specimen is obviously larger than that for the quenched specimen in the two stages within the temperature. Second, the extent of plastic deformation region becomes short at a given temperature by annealing the quenched specimen.

References

[1] Dryden, J. S.; Morimoto, S.; Cook, J. S. *Phil. Mag.* 1965, *12*, 379.
[2] Fleischer, R. L. *J. Appl. Phys.* 1962, *33*, 3504.
[3] Friedel, J. *Dislocations;* Pergamon Press; Oxford, 1964; p 224.
[4] Christian, J. W.; Masters, B. C. *Proc. R. Soc. A* 1964, *281*, 240.
[5] Foreman, A. J. E.; Makin, M. J. *Phil. Mag.* 1966, *14*, 911.
[6] Sumino, K. *Jpn. Inst. Met.* 1972, *11*, 31 (in Japanese).
[7] Alden, T. H. *Trans. Met. Soc. AIME* 1964, *230*, 649.
[8] Davis, L. A.; Gordon, R. B. *J. Appl. Phys.* 1969, *40*, 4507.
[9] Evans, A. G.; Pratt, P. L. *Phil. Mag.* 1970, *21*, 951.
[10] Suszyńska, M. *Kristall. Technik.* 1974, *9*, 1199.
[11] Michalak, J. T. *Acta Metall.* 1965, *13*, 213.
[12] Sprackling, M. T. *The Plastic Deformation of Simple Ionic Crystals*, (edited by Alper, A. M.; Margrave, J. L.; Nowick, A. S.) Academic Press; London, 1976; p 203.

INDEX

A

activation volume for dislocation motion, 21, 198
aggregates (trimers), 46, 198
alkali halide crystal, 198
amplitude dependence of internal friction, 20, 198

B

bending angle of a dislocation pinned by an impurity, 198

C

critical temperature, 7, 8, 32, 45, 52, 53, 56, 64, 84, 105, 107, 135, 143, 169, 183, 199
cross slip, 177, 180, 186, 199

D

direct observation of the dislocation, 19, 199
dislocation density, 9, 11, 19, 20, 22, 25, 26, 58, 165, 170, 175, 177, 180, 182, 186, 194, 199
dislocation multiplication, 8, 165, 172, 173, 177, 199
dislocation velocity, 13, 91, 94, 115, 116, 121, 133, 149, 199
dislocation velocity-effective stress exponent, 8, 83, 90, 93, 95, 99, 102, 105, 112, 118, 120, 122, 199

E

effective stress due to an impurity, 199
elastic compliance, 199
elastic stiffness, 199
electron microscopy, 19, 26, 175, 199
energy for a dislocation motion, 199
etch pit, 26, 175, 199
extent of plastic deformation region, 9, 194, 195, 199

F

flow stress, 7, 11, 12, 13, 15, 16, 19, 20, 44, 46, 47, 58, 94, 108, 136, 142, 146, 147, 155, 165, 172, 199
force acted on a dislocation, 199
force-distance curve (relation), 199
forest dislocation density, 9, 57, 170, 173, 175, 176, 180, 181, 185, 186, 194, 195, 199

H

heat treatment, 9, 19, 25, 42, 105, 119, 120, 187, 190, 195, 199

I

internal friction measurement, 19, 21, 199
internal stress, 13, 19, 20, 22, 54, 94, 141, 143, 199
ionic size, 47, 56, 61, 63, 181, 186, 199

J

jog, 200

L

long-range obstacle, 12, 200

M

microhardness test, 19, 200
mobile dislocation density, 9, 20, 94, 95, 175, 182, 186, 195, 200

P

pinning points, 177, 186, 200
plateau, 7, 9, 27, 32, 33, 34, 35, 43, 46, 47, 57, 74, 86, 98, 109, 121, 137, 138, 140, 142, 143, 150, 154, 166, 169, 170, 175, 183, 185, 187, 189, 192, 195, 200, 201
pre-exponential factor ($\dot{\varepsilon}_0$), 182, 200

proof stress, 19, 200

R

resistance of obstacle to dislocation movement, 200

S

serrated flow, 200
short-range obstacle, 7, 12, 14, 83, 126, 200
solution hardening, 7, 11, 47, 200
strain rate, 12, 13, 19, 22, 61, 144, 149, 171, 187, 192, 200
strain-hardening rate, 9, 193, 194, 195, 200
strain-rate cycling tests associated with the ultrasonic oscillation, 7, 8, 22, 64, 135, 200
strain-rate sensitivity due to aggregates, 200
strain-rate sensitivity due to impurities, 74, 76, 86, 109, 121, 150, 155, 183, 200
strain-rate sensitivity of effective stress, 34, 200
strain-rate sensitivity versus stress decrement curve, 8, 98, 138, 143, 165, 169, 170, 183, 184, 192, 200
stress relaxation test, 19, 200
stress versus strain curve, 176, 200

T

tan δ , 26, 201
tetragonality, 8, 44, 46, 63, 80, 101, 201
thermal fluctuation, 201
three-stage strain hardening, 169, 201

V

vacancy, 11, 80, 201
viscous motion, 201

W

width of force-distance curve, 127, 201

Y

yield stress, 19, 25, 26, 32, 33, 48, 54, 84, 141, 143, 201